住哪？2

区伟勤 著

kitchen

中国建筑工业出版社

图书在版编目(CIP)数据

住哪？2 / 区伟勤著. -- 北京 : 中国建筑工业出版社, 2014.11
ISBN 978-7-112-17431-7

Ⅰ. ①住… Ⅱ. ①区… Ⅲ. ①室内装饰设计－中国－图集②散文集－中国－当代 Ⅳ. ①TU238-64②I267

中国版本图书馆CIP数据核字(2014)第256361号

责任编辑：唐 旭 杨 晓
责任校对：张 颖 王雪竹

住哪？2
区伟勤 著
*
中国建筑工业出版社出版、发行（北京西郊百万庄）
各地新华书店、建筑书店经销
恒美印务(广州)有限公司印刷
*
开本：889×1194毫米 1/20 印张:16 字数：220千字
2014年11月第一版 2014年11月第一次印刷
定价：115.00元
ISBN 978-7-112-17431-7
（26258）

出版说明

本书是作者区伟勤先生继《住哪？》“再”住过的50余个酒店的室内设计记录绘本，每个酒店都配有作者当时的心得体会。由于是即兴所作，手稿中难免存在字迹潦草、语句不通等不当之处。与《住哪？》相同，为保留作者原汁原味的推敲与记录，编辑仅对手稿部分进行细微的调整，而在每篇手稿后面都配有相应的文字说明，是在手稿文字的基础上进行了加工整理的。本书文字内容以此文字说明为准，特此说明。

序

陈文栋

香港陈文栋设计有限公司董事

区伟勤、大师弟，
他称我大师兄，我也就不客气了。
《住哪？2》，
这是一本“旅行札记”，
一本专业设计师“周游列国”，
用心绘画、日积月累、汇集而成的“手稿”。
这种辛苦的“累活”，对于年近“耳顺”的我，早已避而远之，金盘洗手了。
多么渴望再有这种激情，这种态度，这种纯真与虚心……
无时无刻不带着一把自己心爱的卷尺，一支笔，一张纸，把生活中一切自己感到美丽的、实用的、有设计价值的东西记录下来，努力学习，不断积累……
这，就是一个大学毕业生，
一个刚刚走进社会，
迈向事业征途的我们的身影。

Grand Ou, my junior college mate, he call me "senior college mate", so I call him back frankly. The book Where to Live is a travel note of a professional designer who travelled "all over the world". He drew by heart, accumulated day by day, then gathered a book of hand drawing. It's a very hard work for me in nearly sixty years old. I already tried my best to keep such hard work away as far as possible. I desire to have such passion, attitude, sincerity and modesty as Grand... At any time, he takes a tape measure, a pen and paper with himself, records anything he found beautiful, useful or valuable for design, learns and accumulates experience... This is an image of a university graduate, a person who just started his career journey step by step, like us.

还是这一句：
光阴似箭，日月如梭……
多么怀念青春的岁月，
大师兄在职场上奋斗着二十多个年头，
大师弟在市场创业、拼搏、获得了今日的成功。
如今，
一个统领着二百多名员工的设计公司老板，居然仍然有热情，有动力，真心实意地去学习、去坚持、去写写、去画画……
这是一个后来者的楷模，
一个让人肃然起敬的大师弟。

I still say this sentence: Time flies like an arrow… How I miss the days in young ages. I have struggled for my career for over twenty years, while my junior college mate Grand started his business, then worked hard, and got success today. Now, as a boss leading a design company with over two hundred staffs, he even still has passion and power to learn, insist, write and draw sincerely…He is a role model for all young designers, a respectable junior college mate of mine.

呵呵！
我在想：“住哪?”的读者应该是些什么人?
这是一些很有耐性，不惧乏味，
同样揣怀着走向成功的梦想，
刻苦而努力的年轻人吧！

Ho ho! I am thinking: Who is the reader of Where to Live? It should be some young men who have patience, not afraid of dullness, have the same dream for success and work hard.

虽然，
这个年代已经属于“互联网”，
已经属于新一代熟练敲打键盘的年轻人。
但我想，
潇洒的文字，精美的手稿，
娴熟的草图，哲理的日记，
闪光的思想，总应值得让人尊重、佩服。
从中获得希望的热情、深邃的智慧……

Although this is an internet era, which is belong to the youths familiar with keyboard, I think such glamorous words, beautiful drawings, mature layouts, philosophic diaries and blink thoughts are all worth appreciating and admiring. I got the passion of hope and deep wise from this book…

大师兄：陈文栋
Senior collage mate, Venden Chen

4/8/2014.广州星河湾.夜
Night in Star River Guangzhou 4/8/2014

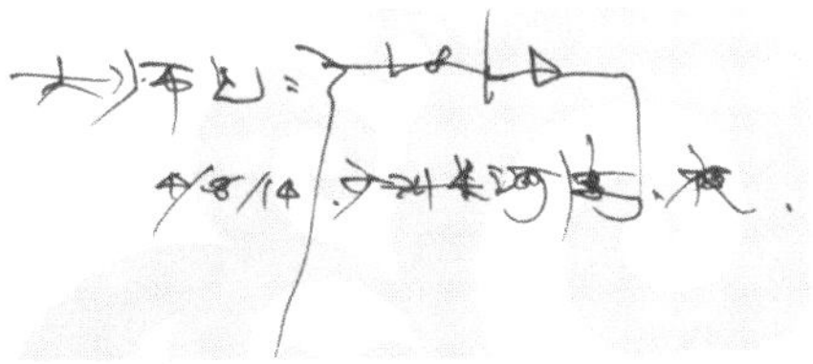

林学明
集美组总裁
中央美术学院城市设计学院教授

不倦的行者

A Traveler Who never be Tired

旅行是学习，是文化体验。把旅途中的所见所闻，学习心得毫无保留地奉献给大众，对于区伟勤来说，实现了他人生价值的一部分，工作于快乐之中。依我看，他的分享，体现的是一种美德……

Travelling is kind of learning and culture experience. Sharing all what he found and what he learned during travelling just realized part of Grand's value of life,Working in happiness. In my opinion, his sharing is also kind of virtue...

或差旅途中，或度假，或考察，不管在哪个城市，到了哪个国家，首先体验的是酒店。在不同的国度，不同的地域，不同的文化背景下，酒店往往成为客人对这个城市的第一印象。

Maybe in business trip, in vocation, or in observation trip, in any city or any country, the primary place is hotel. In different countries, different areas and under different culture, hotel gives the first image of the whole city to customer…

作为一个职业设计师，免不了习惯性地以专业者的眼光，对所住的酒店品头论足，慢慢细品一番。如：酒店的文化品位、主题特征、功能布局、服务动线、客房的平面布局、功能配置的合理性、空间界面的处理、尺度的把控、家具细节的设计、各空间的选材、色彩的搭配、智能化设计、灯光控制设计、艺术品陈设等等。酒店的美誉度取决于酒店的设计和酒店管理。看酒店的设计就能判断一个设计师的功力和生活态度，好的设计往往留给人以难忘的艺术享受。一个好的酒店自然也成为设计师议论的话题；或提出批评，或提出疑问，更多的还是学习。有的还成为引领商业设计风尚的标杆，设计师竞相参考学习，区伟勤的旅行笔记是后者。

As a professional designer, I get used to criticizing each hotel I stay carefully in a professional view, such as the culture level, theme, facilities setting, service, layout of room, rationality of function, interior space division, space size, furniture design, material of interior decoration, color collocation, intelligence design, lights design, display-art, etc.. The hotel design and management decide a hotel's public comment. We can judge a designer's skill level and attitude of life through his design of hotel. The perfect design always impresses customers with its art. And a good hotel is also a topic of designers. They criticize or question about the hotel design, and more usually they learn from it. Some hotel has become the fashion mark of design business, and the learning model for other designers. The hotels mentioned in Grand's book are this kind.

“读万卷书，行万里路。”旅行成为了滋养区伟勤设计观的源泉，他跑遍了祖国大江南北，游历世界各国，到都市、到乡村、到海滨、到沙漠、到草甸、到高原，虚心向大自然学习，向先进发达国家的城市学习。通过对城市文化历史的了解，对不同肤色人群的观察，了解不同阶层的生活习俗，生活方式和价值观，为日后的设计丰富内涵。从生活中品精微，广采博收，这也是他旅行的态度。

"Read ten thousand books, traveling thousands of miles." Travelling became a spring for developing Grand's sense of design. He has been to all around China and many other countries, including city, village, beach, desert, steppes and plateau. He learns from nature, learns from the cities of developed countries. Through comprehending of cities' culture and history, and observation of people with different skin color for knowing their custom, life style and values in different stratums, he develops his design inspiration day by day. Experiencing each detail in life and gathering information, that is his attitude of travelling.

区伟勤把自己对酒店设计的理解一一记录下来，养成习惯并持之以恒十数年。每次下榻新的酒店，便对房间的布局进行测量手绘，及时记下设计心得。无疑，日积月累的学习帮助他的事业走向成功。细看他的手绘和文字，他个性温文儒雅，他特有的细嚼慢咽式的学习态度，值得年轻的设计师学习。

Grand keeps recording his opinion of each hotel for over ten years. Each time staying in a new hotel, he measured and drew the layout of hotel room, and recorded his comment for hotel design. It's no doubt that learning years by years brings success of his career. According to his drawings and words, he is gentle and cultivated and learns each detail carefully. Young designers should learn from him.

2014.10.1

2014年国庆节

National Day, 2014

杨邦胜
YAC杨邦胜酒店设计顾问公司董事长、设计总监

温度
Temperature

首先恭贺伟勤兄“住哪？”已经出到第二部。书中一如既往的手写稿，让人眼前一亮，更让人心生温暖。

First, congratulations for Grand's sequel of Where to Live! Like the first one, all articles are handwritten. That is impressive, and makes me feel warm.

作为一名从业较早的设计工作者，我们也曾经拿笔在纸张上丈量空间，传达创意，那时候画下的每一笔都带着思考，因为一旦出错，就意味着从头再来。

As designers who started this career in early years, we also used to hold a pen to make a plan and express our creation on paper. At that time, we considered each line before drawing, because we had to restart drawing if we made any mistake.

现在电脑软件的方便快捷取代了笔触的随性洒脱，画一张图纸的时间越来越快，图纸也越来越没有感觉，少了些手绘的灵性及手笔连心的温度。

Nowadays, the convenient computer software replaces the casual and free hand drawing. It takes shorter and shorter to finish a drawing, but we feel less and less on the drawing, too. The soul and the temperature in heart of hand drawing are missing.

虽然我依然喜欢随手抓起身边可以涂写的纸张，随性记录身边的点滴，或偶偶迸发的设计灵感，但我不像伟勤兄勤勉，几十家酒店的坚持，认真、执着且有序。

I still like taking some paper nearby to record anything around me, or some inspiration I suddenly got. But I am not as hard working as Grand. He insisted drawing dozens of hotels carefully, tenaciously and orderly.

感谢伟勤兄的坚持，用心的记录，让我们能够站在他的视角，了解那些我们所熟知的酒店的另外一面。每一张印有酒店LOGO的便笺纸，就像一个地标，等待你我去发掘。

Thanks to Grand's persisting and record by heart, we can know the other side of those familiar hotels from his eyes. Each piece of note paper with hotel logo is like a landmark, which is waiting you and me to find.

做酒店设计这么多年，习惯站在客户、市场的角度去反思我们的设计，“住哪？”就像是一种指南，提醒我设计中那些不容忽视的细节及体验。无论时空转换，风格如何演绎，设计者都不能忽略空间中应有的温度。

As a hotel designer for so many years, I am used to reflect our design in customer and market's view. Where to Live is like a guidebook, reminding me the details and experience which can't be ignore. No matter where the hotel is, or what style is designed, the designer mustn't neglect the temperature which the space should have.

所以在我眼里，“住哪？”已经不仅仅是跟酒店相关，体验性、评论性的书籍，它更像一剂良药，让我们出发许久，站在艺术与商业的路口，重温了设计之初的美好。

So in my opinion, Where to Live is not only a book about hotel experience and solution, but also like some effective medicine for us, who have worked for design for too long and mixed art and commerce, helping us to find back the goodliness of pure design.

伟勤是个睿智的人！我喜欢与他交谈，也更爱看他写的书。
Grand is a wise man! I like talking to him, and now I prefer to read his book.

2014.9.10

Contents

目 录

后记

旅行日记

鸣谢

Accumulation

积累

积累

过程很简单，现代人对整理东西皱眉，头痛：相片资料，文字资料，喔喔杂料……反正说是以后再处理，往往干脆就不整理了。历史就历史吧，过去就过去了，分分片段，就好了。或者两年是一个不错的选择，不长不短，当然也是由自身的小时候慢慢培养成的，这多得公司一年四季的春夏秋冬季刊，支撑这做法当年整理所的历程，行踪、见闻等等。

翻出这两年多的往旅，也有十数篇的文章是当时写的，更有几趟的国外旅程留下的日记，文字的印记，这样再整理的压力就不大了，也就更有热情去补充，让它们合序地成文、成章成书了！

积累是靠留下一点点，而不是一蹴而就，正如学习一样，不断重复又重复，温故而知新，没忘记的旧也就不会浮掉了。

当下汕头房地产速度之快不得不让人惊叹。同行、同仁也一样，但，也要向同行学习和交流前进方式：辛勤的是劳动，收获的是经验，也要注意设计、设计过程构成与我们建筑设计的协调，这样才能让我们、你们、他们在不断的影响、交流、碰撞当中得到提高，不经意间，功力提升！

我想也要讲讲学习的，那就是勤奋一点！

Accumulation
积累

过程很简单，现代人对整理东西皱眉，头痛：相片资料、文字资料、理财资料……反正就是积重难返，往往干脆就不整理了，历史就历史吧，过去就过去了，分分片段，就好了。或者两年是一个不错的选择，不长不短，当然也是由每季的小结而慢慢搭成的。这多得公司一年四季的春夏秋冬季刊，每期的专栏用照片被迫当季整理自己的历程，行踪、见闻等等。

The process is quite simple, modern people are headache and frown to sort out so many document such as the photo information, word & literature information, financial information…Anyway, too much means tired only. Some people even simply wouldn't sort out, let history gone with history, past be the past, just took out some section is ok. I think 2-years may be a good choice, not long and not short time, of course it was consist of summary in each season and thus come into being, this thanks to the periodical of Spring, Summer, Autumn, Winter issued by Four Seasons Company. The photo for each column had to sort out my own Journey, trip and whereabouts etc. as per the sequence of each season.

翻出这两年多的住旅，也有十数篇的文章是当时写的，更有几趟的国外旅程每天的日记、文字的印记，这样再整理的压力就不大了，也就更有激情去补充，让它们有序地成文、成章成书了！

Took out the diary recording these two years of living in the hotel, there are several articles that are written at that time, there are diary and words for several trips of journey abroad. Thus there is no pressure on me to sort it out once again, also have more passion to add some new element and fulfilled the content.

积累是靠每天一点点，而不是一蹴而就，正如学习一样，不断重复又重复，温故而知新，设计师的学习也就不会浮躁了。

Accumulation is starting from a little bit but not accomplish at one time. Just like the study, we need to repeat it again and again, recover the old ones and knew the new things, thus the designer won't be impetuous in learning this way.

当下酒店开店速度之快确是难以想象，国外、国内也一样。住，也是向同行学习和致敬的方式。辛勤的是劳动，积累的是经验，也应该让设计过程成为我们财富积累的过程。这样才能让我们、你们、他们在不断的住啊、吃啊、玩啊当中得到提高，不经意的，功力提升！

The opening speed of the hotel is indeed out of imagination nowadays, the same situation at home and abroad. As for living, we also need to learn from our peer and the way to show respect to them. Hardworking is a laboring job, accumulation is the experience. We should let the design process become the process that we accumulated the fortune. In this way, we can continue to live and eat and play and thus improved unconsciously!

积累也是要讲讲策略的，住只是其中一种！

Accumulation is quite particular in terms of strategy, and living is only among of them!

"雨后春笋"

追求GDP的价值也体现在酒店的开业建设上，以前住五星级酒店还担心会住重复的，现在呢，一个城市特别是一线的或旅游城市，其开业的速度真会让你住不过来，比如广州，这两年连续开了四季、文华东方、W、朗豪、接着还有柏悦、康莱德等等，几乎一线品牌都来了，杭州、上海、北京或更厉害，这样看来，设计师可以加入酒店试睡员行列，体验一下免费住不同的新开酒店的"待遇"，也顺便动动脑子写写试住后感，那是不错的兼职啊！

雨后什么长得快？春笋啊！很快就成林了，前提是让人吃掉了大半先，新开的酒店越多，竞争将越激烈，相信也会很快达到半饱和状态，拼品牌、拼服务、拼营销，这就是成熟市场的标志和常态。

春笋长大了，我们就有福了！

追求GDP的结果也体现在酒店的开业速度上，以前住五星级酒店还会担心会住重复的，现在呢，一个城市，特别是一线或旅游城市，其开店的速度会让大家住不过来，就说广州，这两年连续开了四季、文华东方、W、朗豪，接着还有柏悦、康莱德等等。几乎一线品牌都来了，杭州、上海、北京或更厉害，这样看来，设计师可以加入酒店试睡行列，体验一下免费住不同的新开酒店的“待遇”，当然要动动脑子写写试住后感，都是不错的兼职啊！

The result for pursuing GDP is also reflected in the rapid opening speed of hotel. Before we will worry about the Five Star hotel may be repeatedly constructed. However, now one city, especially the first-tier or tourist city, the opening speed is far quicker than we can imagine or the real demand of the people. Take Guangzhou for example, Just these 2 years, many hotels of big brands opened one after another, such as Four Season, Mandarin Oriental, W, Langham, then Bai Yue, Conrad etc.. Almost all the top brands are all coming, let alone the hotels in Hangzhou, Shanghai, Beijing. Seeing in this way, the designer can also join into the rank of trial sleeping in hotel and experience the VIP treatment in different new hotel for free. Of course, you need to use your head and write how you feel when living in this hotel. It is also indeed a very good part-time job!

雨后什么长得快？春笋啊！很快就成竹了，前提是先让人吃掉了大半，新开的酒店越多，竞争得越激烈些，相信也会很快达到半饱和状态，拼品牌，拼服务，拼营销，这就是成熟市场的标志和常态。

After the rain, what grows more quickly? the answer is spring bamboo shoot! We will become bamboo very soon, but the precondition is that we need to be eaten half firstly before making money. However, the new open hotel still be springing up like bamboo shoots after the rain and competing fiercely. I think it will become half saturation state very soon. Competing for brand, for service, for marketing, this is the symbol of a mature market and it became very normal nowadays in today's market.

春笋长大了，我们就有福了。
When the bamboo shoots grow up, we can just enjoy the delicious bamboo shoots.

Whom had You Found?
——To the People Who Writing the Forward

找到了谁？——给写序的人

找到了谁？——给写序的人

找身边的人，了解我的人，这是我的第一感！

出《住哪儿》，一则不求名，不求利，只是为表一下自己这几年行走的足迹，二来也算是对一些酒店，特别是客房的"评论"，片言只语，或者你我也一样有住的活动，不一定把它记住，看到这些"乱七八糟"的印象所画，也许有你也住过的，或共鸣，或气愤，或认同，或吐槽，如果你有这种感觉，那我就找对人了！

住哪儿不重要，关键你落脚的点带给你什么，这也是这本书的期望。期待你看到它、看了它，再度闪起心中隐隐的欲望，拿起笔，而不是打开电脑。

找到了谁，不需名贵的墨水，那是一种舒适的享受！

谢谢你的"举手之劳"，写序的人！

Whom had You Found?—To the People Who Writing the Forward

找到了谁？——给写序的人

找身边的人，了解或不甚了解我的人。

Find the people around you who get to know you or do not know me at all.

出《住哪？2》，一则不求名，不求利，只是发表一下自己这几年行走的足迹，二来也算是对一些酒店，特别是客房的“评论”，片片几言，或者你我也一样有住的经历，不一定把它记住，看到这些“乱七八糟”的即兴所画，也许有你也住过的，或共鸣，或气愤，或认同，或吐槽。如果你有这种感觉，那我就找对人了！

I finally publish the book "Where to live 2". First I need to mention that, neither did I make it for profit, nor the pursuit for fame. The reason I write this actually is only for the purpose of sharing my track and experience of these years. The second reason I write this is for the purpose of commenting on the room service in the hotel although there are just several words or you may share the same living experience like me. You do not have to remember it, however, when you see this chaotic impromptu drawing, you will share my feeling. Maybe you have lived there and share the same feeling or you may be angry or agree or complain. Anyway, if you have such feeling, I think I find the right person!

住哪不重要，关键你落脚的点带给你什么。这也是这本书的期望，期待你看到它，看了它，再度勾起写写画画的欲望，拿起笔，而不是打开电脑。

"Where to live" is not that important, and the most important point is that what it brings you. This is my objective of writing this book. We hope that you can see it very soon. When you finish reading this book, it will help arouse your curiosity of writing or drawing something instead of opening the computer.

找到了谁，不要吝啬你的墨水，那是一种简单的享受。
No matter who you have found, please just write something and never save your ink as that is a kind of simple enjoyment.

谢谢你的“举手之劳”，写序的人！
Thank you for your help in writing me this preface, my dear friend.

LONGCHAMP GARDEN
HOTEL & REAL-ESTATE
融程花园酒店

长沙融程花园酒店

★★★★★

1

LONGCHAMP GARDEN HOTEL,CHANGSHA

Address : No.9 Xiangfu Road,Yuhua District,Changsha, Hunan Province
湖南省长沙市雨花区湘府路9号

Telephone : +(0731) 8877 8888

FAX : +(0731) 8990 9999

Http : //www.lcghotel.com/

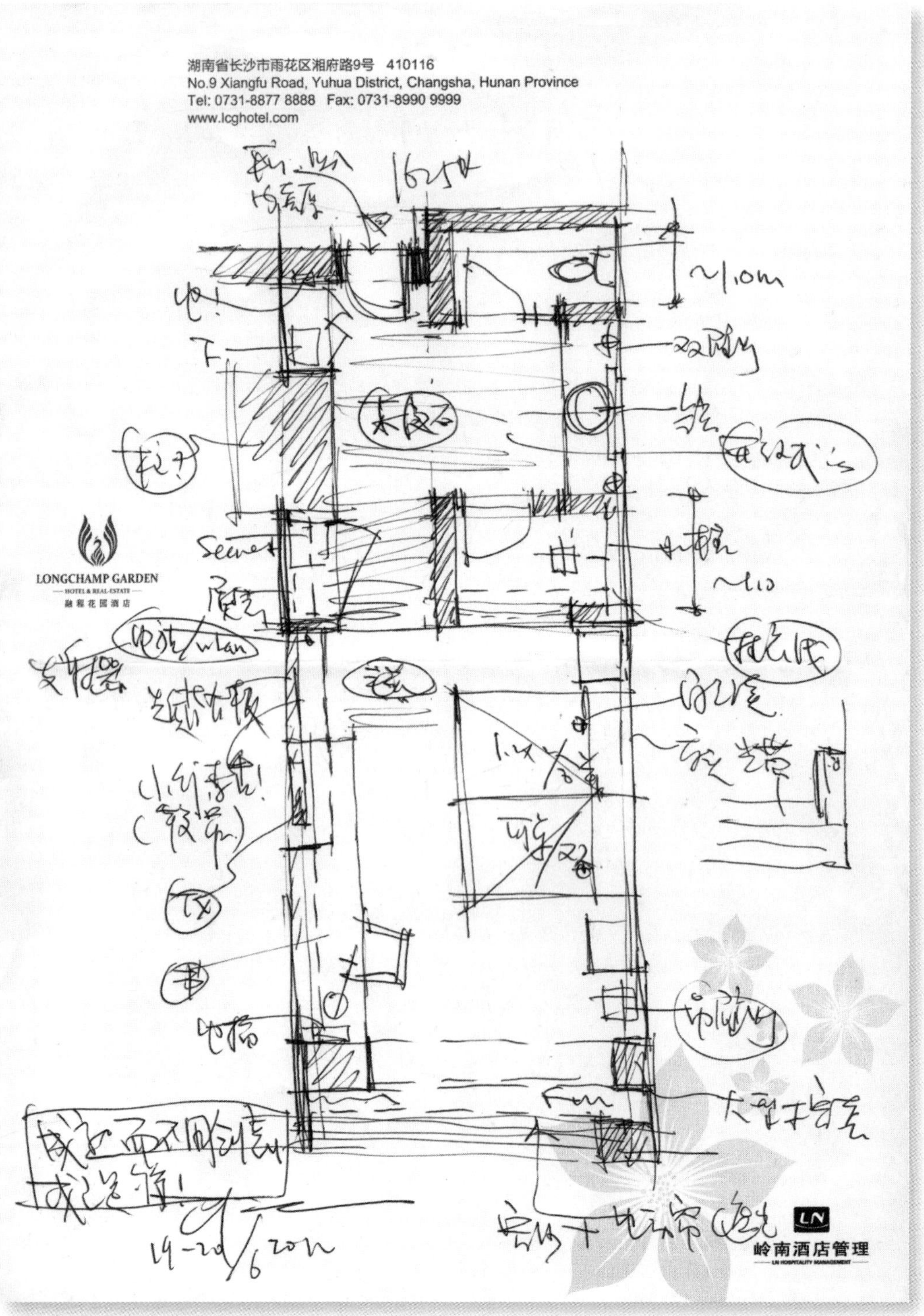

湖南省长沙市雨花区湘府路9号 410116
No.9 Xiangfu Road, Yuhua District, Changsha, Hunan Province
Tel: 0731-8877 8888 Fax: 0731-8990 9999
www.lcghotel.com
LONGCHAMP GARDEN
HOTEL & REAL-ESTATE
融程花园酒店
LN
岭南酒店管理
LN HOSPITALITY MANAGEMENT

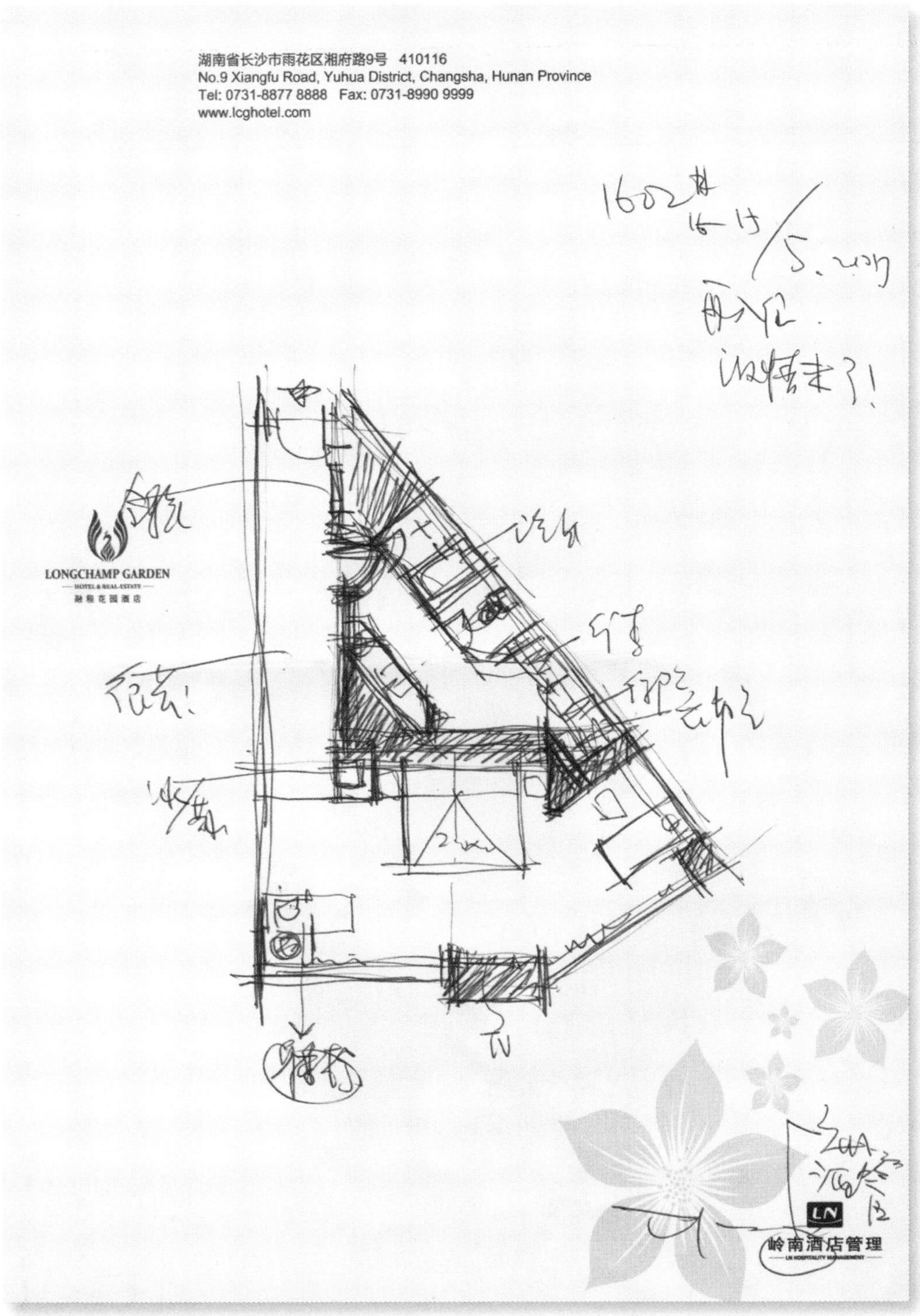

湖南省长沙市雨花区湘府路9号 410116
No.9 Xiangfu Road, Yuhua District, Changsha, Hunan Province
Tel: 0731-8877 8888 Fax: 0731-8990 9999
www.lcghotel.com
LONGCHAMP GARDEN
HOTEL & REAL-ESTATE
融程花园酒店
LN
岭南酒店管理
LN HOSPITALITY MANAGEMENT

成趣的也算创意

第二天的客户开会地点在附近，因此又回到这里入住，长沙融程花园酒店，一家由广州的岭南酒店管理的酒店（与广州的花园酒店、中国大酒店是新中国三大五星级酒店同出一脉），这次还好，住在不同于之前的房型，工整但成趣，惊喜倒没有，但在洗手间所做得这么"开放"也算是一种创新，其实这种做法并不多见，一开始感觉洗手间区域"无门"很不习惯，好像与起居混在一起了。入住倒是有些不一样的体验，厕所是有道门，淋浴给我一不小心洗到"外面"了，心里想这种设计只适合大套房：一人或情侣同住，倒是两位以上来挤在一起体验特色的房，就而得有点尴尬，很可爱。

浅米润、加潘涛一的现代手法还是体现设计中的功力，有时候"成趣者到"也算是一种创意，你说对不？！

Mature Is A Kind of Creativity

成熟的也算创意

第二天的客户开会地点在附近，因此又回到这里入住，长沙融程花园酒店，一家由广州岭南酒店管理的酒店（与广州的花园酒店、中国大酒店及东方宾馆三大元老级酒店同出一脉）。这次还好，住在不同于之前的房型，工整但成熟。惊喜倒没有，但把洗手间做得这么“开放”也算是一种创新，其实这种做法并不多见，一开始感觉洗手间区域“无门”，很不习惯，像与走道混在一起了。入住倒是有些不一样的体验，厕所只有趟门，淋浴结束一不小心就到“外面”了，心里想这种设计只适合大床房：一个人或情侣同住。倒是两张1.2米拼在一起超宽特色的床，床品配得勉强盖得住，很可爱。

The meeting place for tomorrow is arranged at the near place, therefore, we have to move back Longchamp Garden Hotel Changsha here to live. Longchamp Garden Hotel Changsha is one hotel that managed by Guangzhou Lingnan Hotel (It was the same boss with the Guangzhou Garden Hotel and China Hotel and Oriental Restaurant.). For this time is ok, now I can live in the room type that different from before, maturity and in order. All this is nothing special and no surprise. However, I never think that the washing room will be made so open and maybe it is also a kind of innovation. Actually this design is seldom to be seen. When I go into the washing room, I am puzzled why there is no door inside and feel not used to it as you will feel like walking in the aisle. When you lived in and got different experience. There is only one door to cover the toilet.You will be easily exposed yourself outside when you finish taking a shower. I think that this design is only suitable for the room with big beds or it is for living alone or for a pair of sweet couple. The two separate bed of 1.2 meters can be combined together and make it wider to live. I think it is so lovely and the bed items are enough.

浅米调和谐统一的设计手法还是体现设计的功力，有时候“成熟老到”也算是一种创意，你说对不？！

The design of harmonic and unified beige color still embody the power of the design. Sometimes nostalgic old is also a kind of creativity, right?

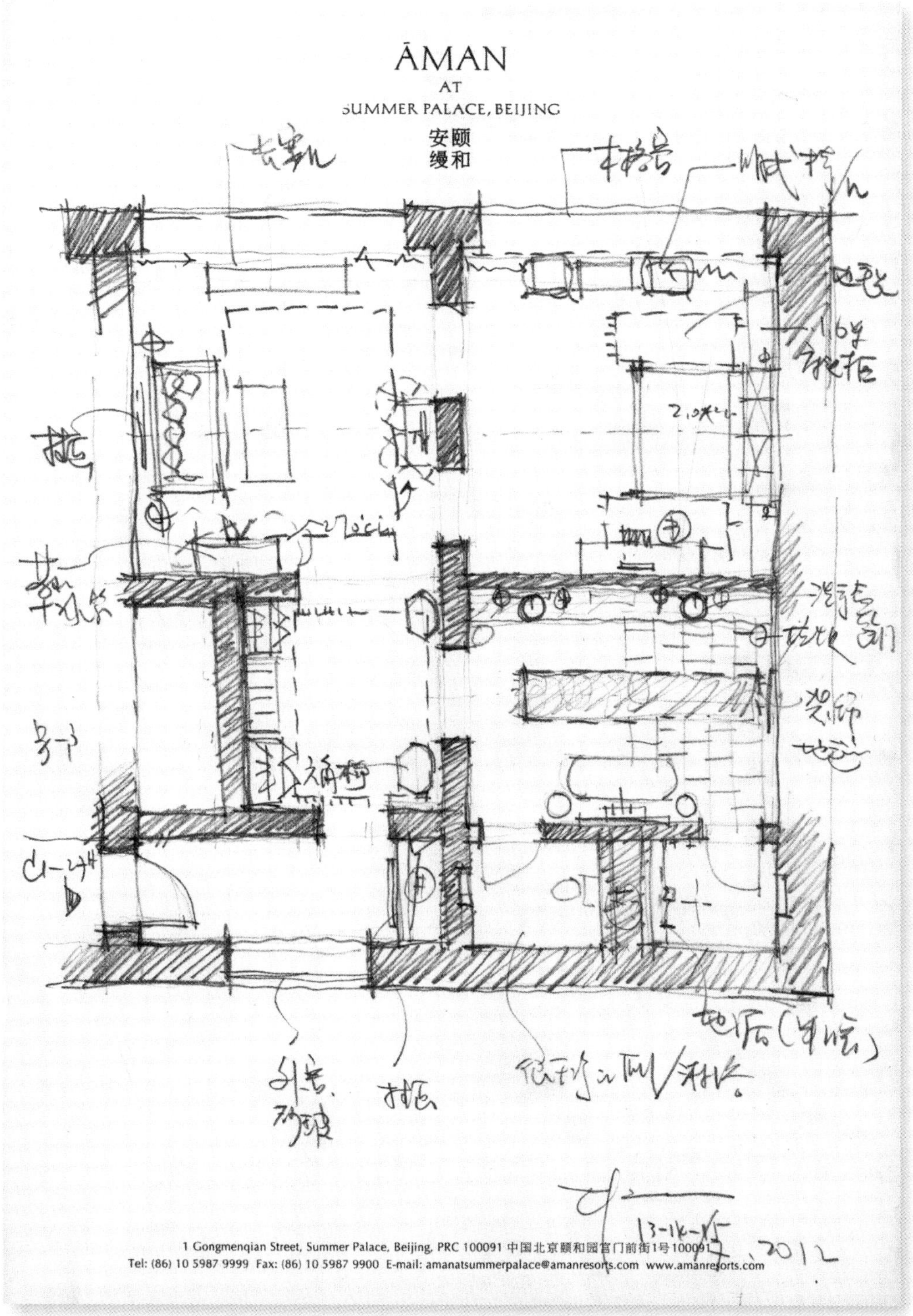
ĀMAN
AT
SUMMER PALACE, BEIJING
安颐
缦和
1 Gongmenqian Street, Summer Palace, Beijing, PRC 100091 中国北京颐和园宫门前街1号100091
Tel: (86) 10 5987 9999 Fax: (86) 10 5987 9900 E-mail: amanatsummerpalace@amanresorts.com www.amanresorts.com

ĀMAN™
AT
SUMMER PALACE, BEIJING
北京颐和安缦酒店
★★★★★

2

AMAN AT SUMMER PALACE, BEIJING

Address : No,1 Gongmenqian Street, Summer Palace, Beijing, 中国北京颐和园宫门前街1号

Telephone : +86-10 5987 9999

Fax : +86-10 5987 9900

E-mail : amanatsummerpalace@amanresorts.com

Http ://www.amanresorts.com/

安缦酒店，你会住吗？

有一些酒店注定只有、只能少数人喜欢，我去安缦到在其中，会让你崇拜！

因为叶怡兰小姐的一系列的书籍，如《隐居·在旅馆》、《终于走到真滋味》都将安缦Amanresort列为必要之首选，于是也想去看看，去住住，下定了决心，于是在国内只有两个选择的情况下去了杭州法云安缦。青竹泥墙，水泥色地砖，灰暗的灯光与极简朴（感觉而已，实质每处细致，包括性护肤极度奢华）让我们追求明快的同行者也一致摇头，细细观察，确实并不感受到它的人气，或者这也是他们的初衷，宁静幸美，据说不能超过70位入住者，否则会降低服务品质，怪不得价格要这么高，这也是一道槛！所以当次入住了也向和享润的西子四季酒店了。或者你不会认同这样的感觉，说去安缦，或者更适合修行者下榻的吧，雨天或泥泞更是凄迷，心情不好奇怎么样！倒是入住的北京颐和安缦非常喜欢了。土吧！一个是整体感觉不错，古色地很精致，对称、大而散落，从而带有古意境，或这是个人偏好：喜庆了。酒店还是吉祥一点好！不然待久了会让人心沉下来，或者宁静也是一部分来自人所追求的。总之安缦也有机会的Fans，而我只喜欢在那里喝喝茶，吃吃喜！

北京的颐和安缦，客房位于二层，管家和与精致到位，导入式的布局，尽显仪式感和意境，拐角、收放，正一下平面就能看出推敲的功力，管家的用材和造型，一具在润光进的两层设备，也注意到长廊下垂的竹卷帘，让文化院门光秀之地溢进来，当然后来到那是报道才进一步理解了设计师的用心良苦，除却只是一面，而是要营造一种"曲径通花"的情趣，管家的手法，意想不到的效果。

客笑：静还是静，这个体验是喜欢的，房间的水果、小甜点也真不错，印象也很深

当然吸引我的还有机会的多，所在是其中

之一，大型的宽摆式专业影院，下沉甲板的大型夜店会所，内藏回式的甲板餐厅，大片花园下午茶，spa等等，一切的一切，都让你够爽！

这样咱们的酒店，你会心动了！

Aman Hotel, will You Live There?

安缦酒店，你会住吗？

有一些酒店注定只有，只能少数人喜欢。或者安缦列在其中，会让你费解！

Some hotels are destined to be loved by few people. Aman Hotel is among of them and you are just difficult to understand!

因为叶怡兰小姐的一系列书籍，如《隐居·在旅馆》、《终于尝到真滋味》都将安缦Amanresort 列为必要的首选，于是也想去看看，去住住。下定了决心，于是在国内只有两个选择的情况下先看了杭州法云安缦：青竹泥墙，水泥色地砖，灰暗的灯光与极简朴（感觉而已，实质每处细致，包括维护都极度奢华）让我们追求明快的同行者也一致摇头。细细观察，确实并不能感受到它的人气，或者这正是他们的初衷；宁静柔美。据说不能超过70%入住率，否则会降低服务品质，怪不得价格要这么高，这也是一道槛！所以当次入住了热闹和柔润的西子四季酒店了。或者你不会认同这样的感觉，法云安缦，或者更适合骄阳当下的日子，雨天或浪漫，更是凄迷，心情不好者忌住！倒是之后入住的北京颐和安缦就喜欢多了，土吧！一个是皇家感觉不错，当然地段特殊，规模大而散落，明媚而富有意境，或这只是个人偏好：喜庆多了，酒店还是吉祥一些好！不然待久了会让你心沉下来，或者宁静也是一部分都市人所追求的，法云安缦也有相当的Fans，而我只喜欢在那喝喝茶，吃吃素！

北京颐和安缦酒店 *Aman at Summer Palace, Beijing*

Due to a series of books published by Ms.Yilan Yip, such as "Elusive life-In the hotel", "Get the real taste finally", both of which listed Aman Resort as the top option. Then I go to have a look at as well and make decision to live there. There are only two hotels available in China. First of all, I went to Hangzhou Fa yun Aman Hotel, which is built up with green bamboo and earth wall, the earth color of bricks, dark light and extremely simple (just my feeling, actually every detail is quite particular , moreover, maintenance is also quite extravagant and luxurious), which makes our peer quite shocked with big surprising. If we watched it carefully, indeed we didn't feel its popularity or it is just what they mean at that time; It is peaceful, soft and beautiful. It was said that the occupancy can not be more than 70%, otherwise the service quality will be ruined. No wonder their price will be so high. This is also the bottleneck for them. Therefore, this time I lived in the bustling and humid Xizi Four Season Hotel. Maybe you won't recognize such kind of feeling. Fayun Aman is more suitable for living in the strong sunny day or in the romantic raining day. You will feel it charming with beauty. I'd suggest you do not live here if you are not happy. After I lived in Aman at Summer Palace, Beijing,I loved them more and more. Am I boorish? One is very nice imperial hotel, of course the road section is quite particular, large-scale but scatter, beautiful and poetic, or maybe this is the individual preference only. It looked more exciting and I liked hotel with more auspicious! Otherwise you would feel heavy if you stayed for a long time. Maybe the peaceful mood is some of the metropolitan people have been pursuing. Fayun Aman also has a group of fans, for me, I only enjoy having some tea there or to be a vegetarian there.

北京颐和安缦酒店 *Aman at Summer Palace,Beijing*

北京的颐和安缦，客房位于二层，简单的步梯到达，步入式的布局，尽显仪式感和意境，扬抑、收放，画一下平面就能看出推敲的功力。简单的用材和造型，一贯低调先进配套设备，也注意到长期下垂的竹卷帘，让庭院阳光柔柔地洒进来。当然后来看了相关报道才进一步理解设计师用心良苦，隐私只是一面，而是要营造一种“雾里看花”的情趣，简单的手法，意想不到的效果。

Aman at Summer Palace, Beijing, the guestroom is located at the second floor, just walk by stairs and the layout of walking type reveals completely the sense of rite and atmosphere, depressed and released, just draw some pictures and layout on the plane and it will easily see the powerful strength for designing. The simple material and shape, the low-key profile and advanced equipment, you also need to pay more attention to the drooping bamboo curtain, let the sunshine project softly from the courtyard. Only when we see the related report later can we know and understand the hard-working job. Private is one issue, but the most important thing is creating one kind of temperament and interest of ambiguous effect like "seeing flower in the mist", simple way but unexpected result.

窃笑，高手还是高手，这个你肯定喜欢的，房间的水果，小甜点也真不错，印象也颇深。

Snickering, master is always the master, you will definitely like this. The fruit in the room, the dessert is really nice and leave me very good impression.

当然吸引我的还有很多。配套是其中之一，大型的点播式专业影院，下沉采光的大型健身会所，内庭园式的早餐厅，大后花园下午茶、spa等等，一切的一切，都让你驻足！

Of course there are so many other things attracting me so much. Among of them, it is the matching equipment. Large-scale on-demand professional cinema, large-scale gym fitness club with heavy and dark light, inner garden style dinning room for breakfast, large garden for afternoon tea, spa, etc., all in all, you will be attracted to stop for a while and appreciate the beautiful landscape.

这样吸引的酒店，你会住吗？！
Such attractive hotel, will you live?

北京颐和安缦酒店 *Aman at Summer Palace, Beijing*

呼伦贝尔蒙古包

3 ★★★★★

GER, HULUNBEIER

Address　：呼伦贝尔

Telephone：不详

FAX　　：不详

Http　　：不详

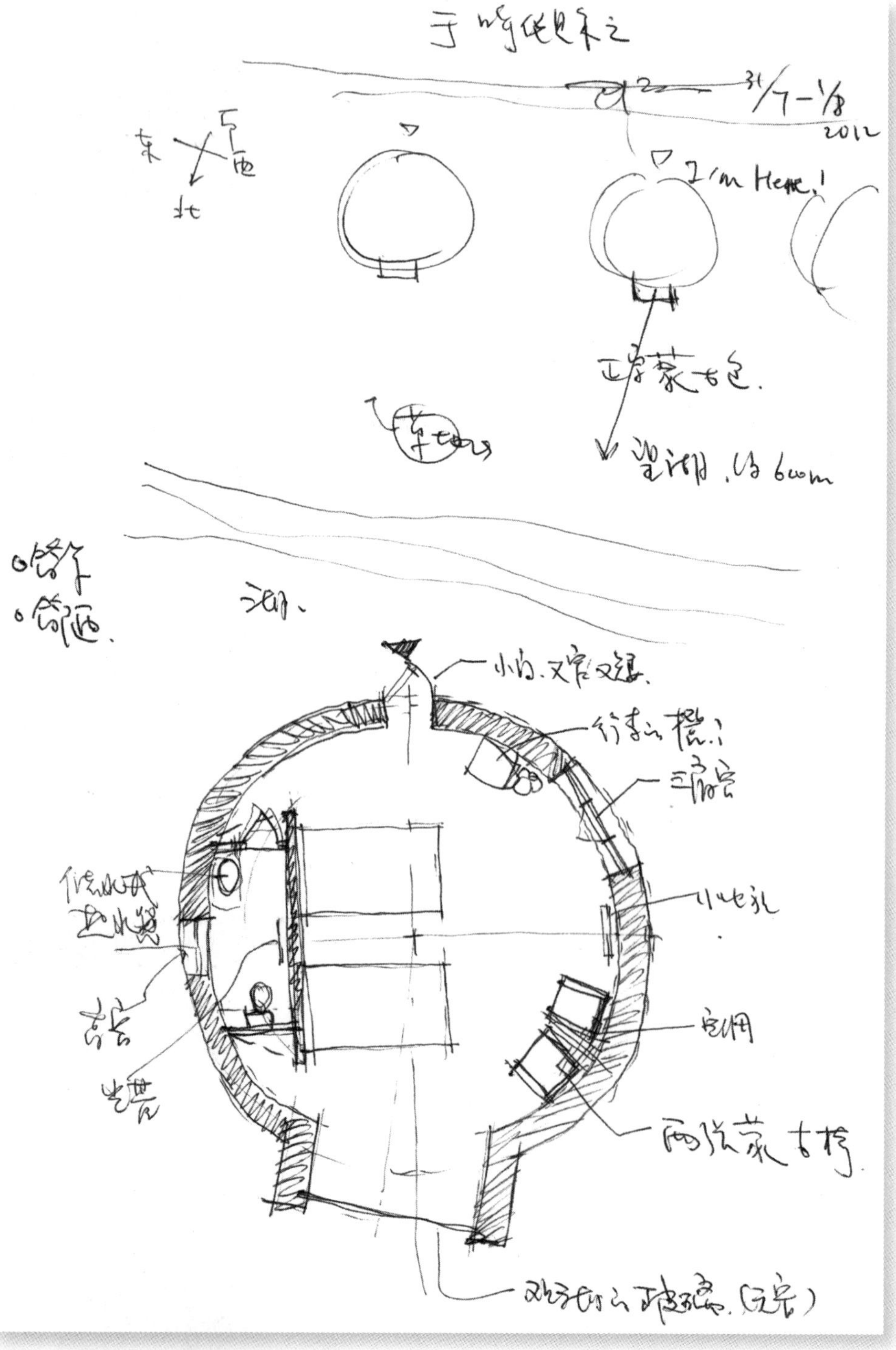

31/7-1/8
2012
I'm Here!

第一次的蒙古包

以前外出，偏于也带相机，到近一两年几乎都是手机拍照摄影了。有些食物手颤抖，喝酒后心颤抖，行走中身体在颤抖，总之都是质量一般般，只能作为自我回忆不可细看，但这都是当时的心情，当时的美景。

第一次去呼伦贝尔大草原，一群来自五湖四海的清华同学，他们这种类似MBA班的意义除了上课就是吃、喝、玩、乐，结交朋友、交流平台，当然有当地的班同学的操心，精心准备的一周的行程，但总是让我们这些懒人尽享各地美食、美酒、美景，以致于对着这么漂亮的湖边的蒙古包房子，印象模糊，匆忙入住，醉醺醺的回来，第二天太阳晒到屁股就被狠狠地拖起了！

回想前一晚的蒙古包大餐，羊牛满桌，美酒当道，欢声笑语，当地更有热情的民族歌舞助兴，以我的小酒量，很快就不行了！

地方人的热情，真让人受不了，但非常尽兴，也算第一次体验了原生态，两宿在大湖边的蒙古包了！

Stay in Mongolian Yurt for the First Time

第一次的蒙古包

以前外出，偶尔也带相机。到近一两年几乎都是手机拍的相片了。有兴奋时手颤抖，喝酒后心颤抖，行走中身体在颤抖，总之相片质量一般般，只能作为附图方式，不可细看，但这都是当时的心情，当时的美景。

I used to take camera sometimes for my trip. In recent two years, almost all photos were taken by mobile phone. But sometimes I couldn't hold my mobile still because of excitement, being drunk or in walking. The photos are not very good. They can be attachment for my articles only, not for appreciation, but all showed my real mood and the beautiful scenery at that time.

第一次去呼伦贝尔大草原，一群来自"五湖四海"的清华同学。似乎这种类MBA班的主题除了上课，就是吃、喝、玩、乐，纯粹的交友、交流平台，当然有当地的崔同学的操心，精心设计的近一周行程，倒是让我们这些懒人尽享各地美食、美酒、美景，以至于对着这么漂亮的湖边的蒙古包房子，印象模糊、匆忙入住，醉醺醺归来，第二天太阳晒到屁股就狼狈撤退了！

I went to the Hulunbuir Pasture Land for the first time, with my classmates of Tsinghua University from all over China. Except having courses, the topic of such MBA class seems to be eating, drinking, playing, and making friends. With the local classmate Cui's careful arrangement, the trip plan for nearly one week was very good. We enjoyed the delicious food, good wine and beautiful scenery in each place, but ignored such beautiful Mongolian yurts by the lake. We just stayed there for short time, even being drunk, slept until the sun was high, and left in hurry!

回想前一晚的蒙古包大餐，羊、牛满桌，美酒当道，欢声笑语，当然更有热情的民族歌舞助兴，以我的小酒量，很快就不行了！

Memory back to the Mongolian dinner party, there were full of mutton and beef on table. We drank, we laughed, we chatted, and there was also ethic dance and song for entertainment. With my small drinking capacity, I was drunk in quickly!

北方人的热情，真让人受不了，但非常尽兴，也算第一次住了原生态、面朝高原大湖的蒙古包了！

The northern people are too ebullient for me, but it made me exciting, too. It was my first time to stay in the natural Mongolian yurt which was facing the big lake of plateau.

呼伦贝尔蒙古包 *Ger ,Hulunbeier*

hotel INDIGO
SHANGHAI ON THE BUND

上海外滩英迪格酒店

★★★★★

4

HOTEL INDIGO SHANGHAI ON THE BUND, SHANGHAI

Address : No.585 Zhong Shan Dong Er Road, Huangpu District, Shanghai , People's Republic Of China
中国大陆 上海市 黄浦区 中山东二路585号

Telephone : +86-21 3302 9999

Http : //www.ihg.com/hotelindigo/

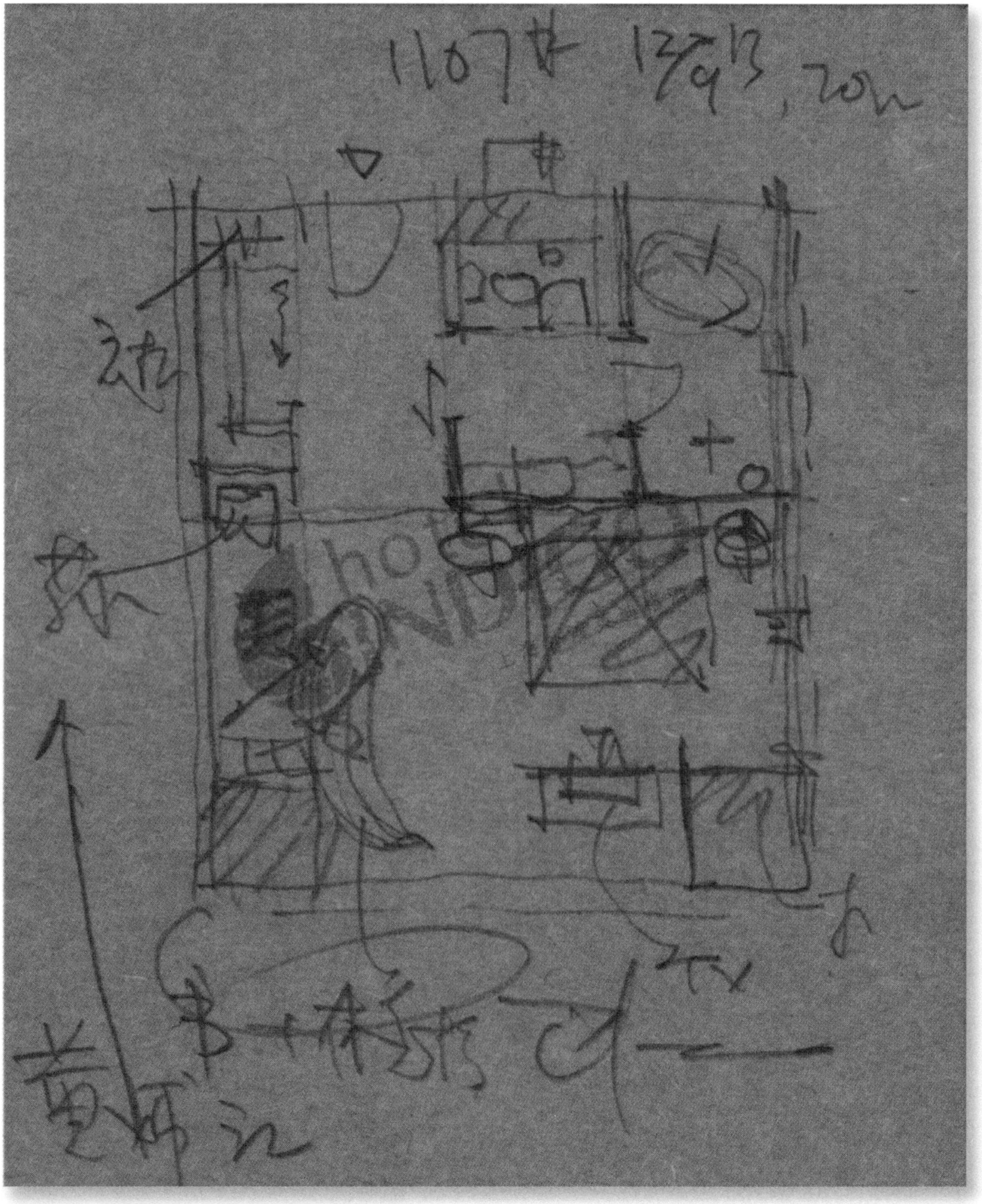

好色之途"

开始不了解洲际旗下的indigo酒店，因为这是国内第一间，位置相当好，外滩边上，带着尝试的心情选择。一到大堂就被艺术了一下，原来和我几年前的设计团队合作几分相似，当然相似的还有一般的工艺和效果，倒是喷漆的调子，灯光的配合艺术家的作品突显了这个风格。后来查了一下英迪格各地的酒店，都是这个整蛊惊吓人的调调！

到前台后，到客房都会让你惊艳大开，大江大河的，或者给我们带来了的"创造潮"的文化的商旅人士的一种精神，心里想更适合年轻一代。早餐倒是一流，黄埔江景一览无遗，大大的浴缸更是吸引，长时间的泡澡，洗去一天的疲劳（泡澡的香木器具，可以泡澡）。

还是一个不错的"好色之途"，大师HBA的作品还是相当可圈可点的，只是挤满了各式"花样"，让你"不得安宁"！

上海外滩英迪格酒店 *Hotel Indigo, Shanghai on the Bund*

Colorful "Tour"
好色之"途"

开始不了解洲际旗下的indigo酒店，因为这是国内第一间，位置相当好，外滩边上，带着尝尝新的心情选择。一到大堂就被"艺术"了一下，居然和我几年前的设计周展位有几分相似，当然相似的还有一般般的工艺和效果，倒是暗暗的灯光，配合艺术家的作品定调了这个品牌。后来查了一下英迪格各地的酒店，都是这个惊喜、吓人的调调！

At the beginning, I didn't know that Hotel Indigo is one of the brands that subordinated to the Intercontinental Hotel as this is the first one at home. The location is quite good and close to the beach, I select it with a good mood of trying something new. When I came to the lobby and I am quite moved by the art. It is quite surprising that it has somewhat like the booth that I had designed several years ago. Of course, the similar one is that the ordinary and common art and effect. The works and tone that the artist had set with the dark light, which together make this brand. Later I checked Hotel Indigo in each place and found the same tone. I think they want to keep this scary tone intentionally and surprise its customers in this way.

上海外滩英迪格酒店 *Hotel Indigo, Shanghai on the Bund*

到标准层，到客房都会让你难挡亢奋。大红大绿的，或者给我们待惯了以“舒适温馨”为主线的商旅人士以一种提神。心里想，更适合年轻一代。景色倒是一流，黄浦江景一览无遗，大大的洗手间更是吸引人。长时间的淋浴，洗去一天的疲劳（淋浴间有小凳子，可以坐浴）。

When you reach the standard floor, the guestroom will make you exciting. Strong red and strong green, which gave us businessman a kind of refreshing feel that quite different from we had used to leading a comfortable office life. I just thought that it may be more suitable to the younger generation. The first-class Huangpu river view completely appeared before your eyes, big washing room is even so attractive. Longer time of showering helps wash a day's tiresome (there are small bench in the shower room, and you can sit down taking a shower).

还是一个不错的好色之“途”，大师HBA的作品还是相当可圈可点的，只是挤满了各种“花样”，让你“不得安宁”！

It is a tour which is full of strong color and the works from the master HBA is quite good and excellent but just filled with all kinds of "patterns" and you can not stay calm any more.

香港W酒店

★★★★★

5

W HOTEL,HONGKONG

Address : Austin Road West Kowloon Station,Kowloon,Hong Kong
香港九龙柯士甸道西1号港铁九龙站

Telephone : +852 37172222

Http : //www.starwoodhotels.com/

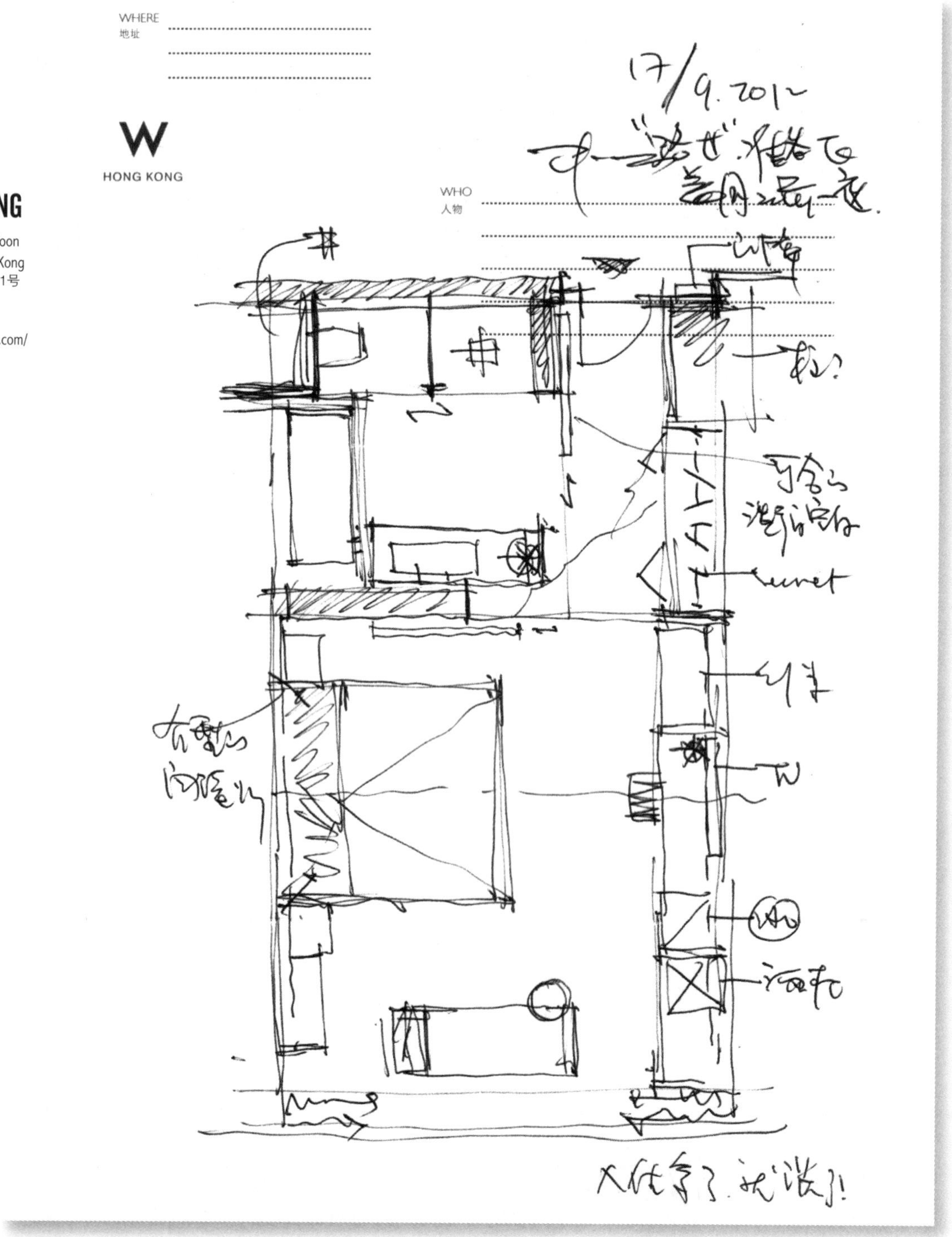
WHERE
地址
W
HONG KONG
WHO
人物
17/9.2012

w hong kong
香港九龍柯士甸道西1號港鐵九龍站
1 austin road west
kowloon station, kowloon, hong kong
+852 3717 2222

W
HONG KONG

路过

09/7, 2012.

再次选择"W"酒店，因为住过，所以仅是为了就能到达香港机场的登机大堂处，第二天一大早就能在这里坐机场快线，便捷地到达香港机场。可惜"路过"首选的酒店。

这几年"路过"所住的酒店并不多，然要能住一天两天的商务旅游式几天的度假休憩；能推荐"路过"的酒店也是有其先天的地理位置和合适的价格比：住得舒适，而不能太贵。香港W酒店只能算不太贵，舒适性倒不是很够，也许只是我们这种住多了的"比较者"的判断，相信喜欢时尚的人还是挺推崇这样的W酒店的。

"路过"也是一种情怀，只需作选择了！

再次选择“W”酒店，因为住这里，可以仅走几步就能到达香港机场的登机办理处，第二天一大早就能在这里坐机场快线，便捷地到达香港机场，可谓“路过”首选的酒店。

I chose W Hotel again just because it's only a few steps away from the boarding desk of Hongkong airport, and the next morning I can take airport bus from here to Hongkong airport conveniently. This is the best choice of hotel on journey.

这几年“路过”而住的酒店并不多，多是安稳地住一天、两天的商务差旅或三几天的度假小憩。能担任“路过”酒店也是有其先天的地理优势和合适的性价比：住得舒适，而不能太贵。香港W酒店只能算不太贵，舒适性倒不是很够。也许只是我们这种住多了的“比较者”的判断，相信喜欢时尚的人还是挺推崇这样的W酒店的。

These years I seldom stayed in hotel in my journey. Usually I stayed for one to two days in business trip, or for several days' vacation. The hotel for journey must have good position and reasonable price. It should be comfortable but can't be too expensive. W Hotel in Hongkong is not very expensive but not comfortable enough, either. Maybe only the guests who used to stay in many hotels have such attitude, like me. I think the people who like fashion would appreciate such W Hotel.

kitchen

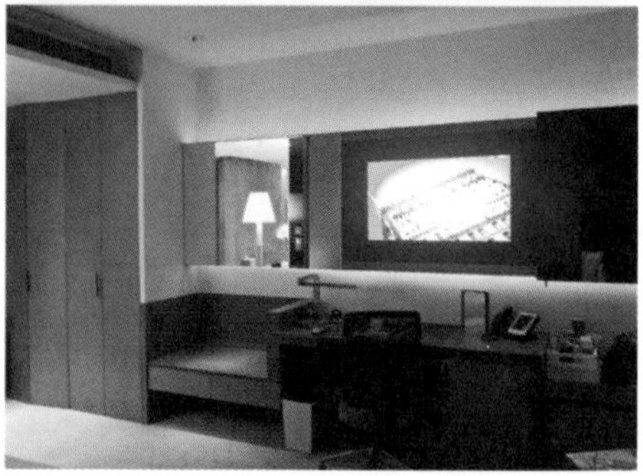

“路过”也是一种生意，只要你选择了！
Hotel for journey is also good for business, as long as you choose it!

香港W酒店 *W Hotel, Hongkong*

WALDORF ASTORIA
HOTELS & RESORTS

纽约华尔道夫酒店

★★★★

6

THE WALDORF ASTORIA HOTEL,NEW YORK

Address : No.301 park avenue new york 10022-6897
Telephone : +212 355 3000
Fax : +212 872 7272
Http : //waldorfastoria3.hilton.com/

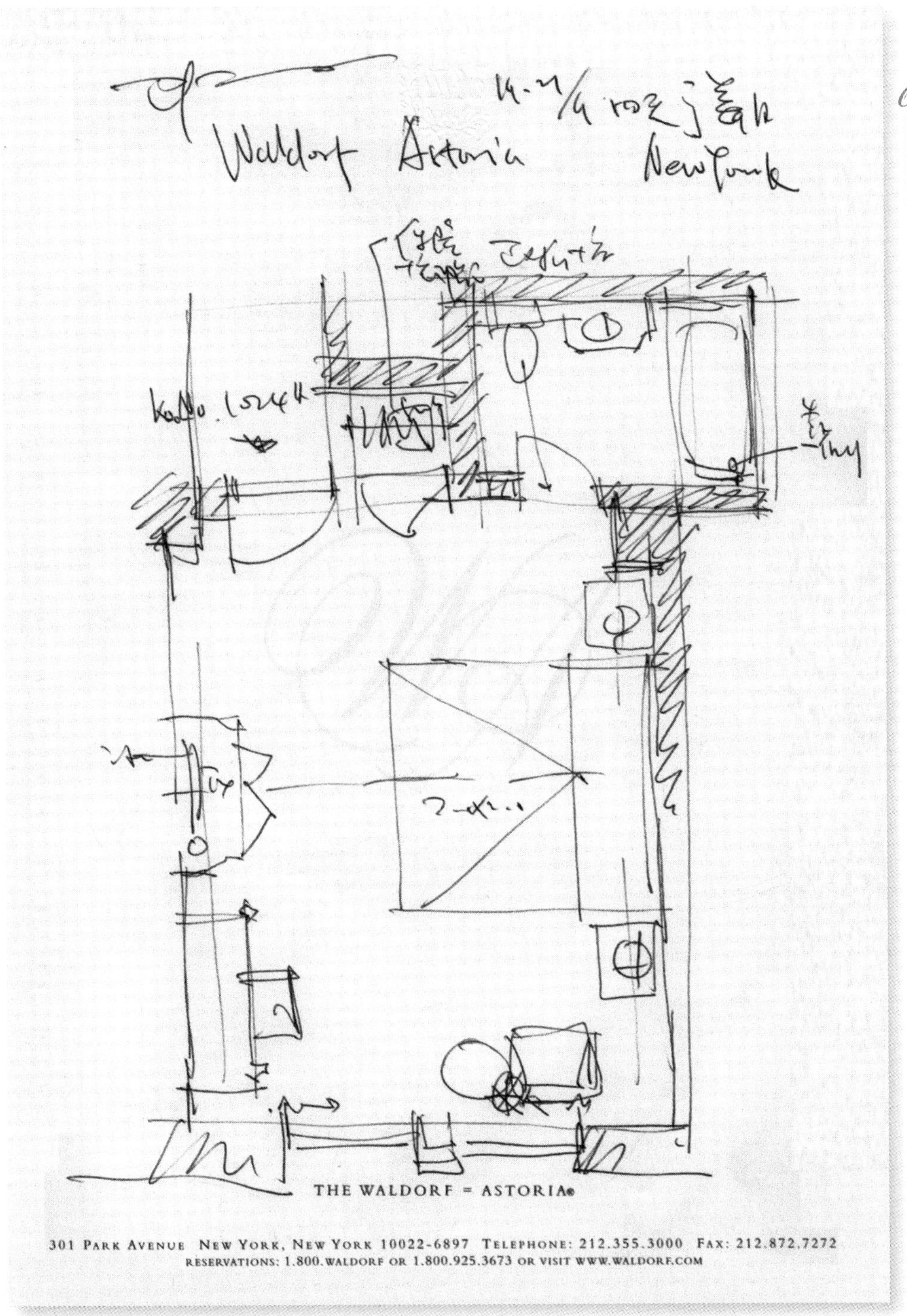

Waldorf Astoria
New York
THE WALDORF = ASTORIA®
301 PARK AVENUE NEW YORK, NEW YORK 10022-6897 TELEPHONE: 212.355.3000 FAX: 212.872.7272
RESERVATIONS: 1.800.WALDORF OR 1.800.925.3673 OR VISIT WWW.WALDORF.COM

奥巴马住哪里？

纽约是这次二十天美国东西岸半自由行的第一站，本来的当天给人感觉是：人多、地脏，选择落脚点位于柏克大道（park Avenue）上经典古老的华尔道夫酒店（1931年建成），回国后翻了一下专业书，才知道它也是Historic Hotel of American其中一员。

不慌不忙在车上到处张望，穿过大半个城市有初始的认识，进入中心区，突然发现越来越热闹，车速慢慢下来，等接近酒店一两个街区，居然大堵，酒店周围水泄不通，也不让靠近，三个方向三两个警察，还有电视台采访直播车守候，向警察打听，原来是奥巴马准备从华尔道夫离开，啊，居然让我碰上了!!

每一个华尔道夫，也是总统下榻的地方，看来我们也选对了。Oh 奥巴马你们、我们也住这，当然房间不同啦！

气派的大堂，深沉的木色，精美金色的藻井，灯柱，古老的摆件，中间1893年的“大笨钟”更是镇店之宝，当然还有热闹的气氛。等候check in的人们，可见要观光的和像我们一样慕名入住的真不少，但是入到房间精木，到处都是掉漆的。可能是当年旧建筑的局限吧。但美式经典，还好，水压很棒，大大的热水澡后，倒头就睡，倒了时差，相信肯定比奥巴马睡得香！

Where does Obama Live?

奥巴马住哪里？

纽约是这次二十天美国东西岸半自由行的第一站，小雨的当天给人感觉是：人少地脏。选择落脚点位于林荫大道（Park Avenue）上经典古老的华尔道夫酒店（1931年建成）。回酒店翻了一下专业书，才知道这也是Historic Hotel of American其中一员。

New York is the first leg of my twenty day's Semi-free trip on the east and west coast in the U.S. Raining give that day the feeling is that few people and dirty land. I choose one old and classic hotel with a name Waldorf Astoria (established in 1931) which is located in the Park Avenue. I came back hotel to check the professional book and know it was also one member of Historic Hotel of American.

EMERGENCY

习惯性地在车上到处张望，希望对这个城市有初始的认识。进入中心区曼哈顿就感受到热闹，车速缓慢下来。等接近酒店一两个街区，居然大堵，酒店周围水泄不通，也不让靠近。三几步就三两个警察，还有电视台采访直播车守候。向警察打听，原来是奥巴马准备从华尔道夫离开。啊，居然让我碰上了！

I had formed a habit of watching all around on the car, hoping to have an initial understanding of this city. When entering into the City Center Manhattan, I started to feel crowded and the car speed will slow down as well. When reaching one or two block from the hotel ,we met a great traffic jam. The hotel was besieged by hundreds of thousands of cars and we could not get close to the hotel. Just several steps, there were two or three polices there, also the TV Interview Car stayed there and enquired from the police. Well, I was told that President Obama prepared to retreat from Waldorf, My God! It was caught by me.

AMERICAN SATELLITE
UPLINK
HD

纽约华尔道夫酒店 *The Waldorf Astoria Hotel, New York*

好一个华尔道夫，也是总统下榻的地方，看来我们也选对了。奥巴马住的，我们也住住，当然房间不同啦！

Well, What a Waldorf! It had become the staying place for the President. It seemed that we got the right option. Well, Obama can live there, of course we like to live, and of course our room will be different!

气派的大堂、深沉的本色、精美金色的藻井、灯棚、古老的摆件，中间1893年的“大笨钟”更是镇店之宝，当然还有热闹的气氛，等候check in 的人们，可见像我们一样慕名入住的真不少。倒是感觉房间特小，到处都是挤挤的，可能是经年旧建筑的局限吧，是纯美式经典。还好，水压很棒，大大的热水澡后，倒头就睡，倒倒时差，相信肯定比奥巴马睡得香！

The magnificent lobby with deep color, the exquisite golden caisson, lantern, ancient ornament, among of them, the “Big Bell” in 1893 is also the treasure of the hotel. Of course the atmosphere is quite exciting. People are there waiting for check-in. You will see it is the most welcomed hotel and people like me only for the famous brand of this hotel. When entering into the room, I find it is quite small and everywhere is so crowded. Maybe it is the limitation that left from years of architecture and it is one pure American style. It is quite good, water pressure is fairy good. After taking a hot shower, I felt into sleep. I was confident that I would sleep better than Obama!

"等级"服务初体验

偶尔在国内航线上也体验一下头等舱的服务，我把分兑换成"逗被"升舱的，不会thankful的越洋还是第一次，这次乘的美国至美国境内的六七次飞行同一选用美国联合航空，标准流程：坐下给饮料，分发到毛毯，换袜子，点餐，有全程的影视媒体及娱乐，自带的ipad根本派不上用场，除了想听听国语的电影。

但吃饭来了，空姐（个个上了年纪的空姐大妈）非常热情地嘘寒问暖，问这问那的，我的听力又不怎么样，点餐时汤料她说"sure"（又不是"soup"），听了半天都难理解，但有一样可让我难忘和感动：飞机平稳后，去了洗手间，顺便擤了一下鼻子，服务我们的大妈过了一会就送来了一盒纸巾和用毛巾包裹住的一大个可乐瓶子热水，塞在我的脖子上。我开始还没反应过来，忙说"Thank you very much."之后更多次问候并换了一次热水！深深体会了等级服务的无微不至。

其实沟通也是一种服务，并不一定只是丰富的创意，好的笔墨，认识的态度和执着的精神都会以作品来持续的回报，相信较会能发浮潜此道，是值得尊重和学习的一个画世，让我常常地记得她们的好。

什么是经典与价值，不可超越的是时间，经典与价值只能用时间去承载和传播，我们不妨多一些去消费这种"奢侈"服务，在享受的同时也要多多支持提升自己和画世的品质，那就值回票价了！

Initial Experience of "Rating" Service
"等级"服务初体验

偶尔在国内航线上体验一下头等舱的服务，或积分兑换或"被迫"升舱的。长途十几小时的越洋还是第一次，这次来回美国及美国境内的六七次飞行同一选用美国联合航空。标准流程：坐下有饮料，有报纸杂志，换袜子，点餐，有全程的影视媒体及娱乐，自带的IPAD根本派不上用场，除了想听听国语的电影。

Occasionally experience the Business Class Service of the Civil Airline or exchange points or "forced" for upgrading the class. For me, it is still the first time to fly across the ocean and go abroad after more than ten hour's flight. This time I choose American United Airline as the sole airline to fly back and forth the U.S. and also some places in U.S. for 6 or 7 times. The Standard service workflow is like this: Drinks are available when you sit down, magazines are available, change the sock and order my meal, film and television media and entertainment program keep non-stopped playing. You will find IPAD is no need to use except that you want to listen to some Mandarin film.

但麻烦来了，空妈（个个上了年纪的空姐大妈）非常热情，口音也重，问这问那的，我的听力又不怎么样，点餐时汤粉她发"sure"（又不是"soup"），听了半天都难理解，但有一样确令我难忘和感动：飞机平飞后，去了洗手间，顺便擤了一下鼻子，服务我们的大妈过了一会就送来了一盒纸巾和用毛巾包裹住的一大个可乐瓶子热水，垫在我的脖子上。我开始还没反应过来，忙说"Thank you very much"，之后更多次问候并换了一次热水！渐渐体会了等级服务的无微不至。

However, trouble is still coming. The airline stewardess or we call airline aunt (alluding to the women stewardess who is old in age) is quite enthusiastic with heavy accent. She kept asking you for this and for that. I have poor listening for English. When ordering food, she always pronounce "soup" as "sure". I have to take great pains in understanding what on earth she means. But there was one thing that made me absolutely memorable and moved. When the plane flied stably, I went to the washing room to blow my nose. The stewardess aunt came over and passed us one box of napkin and one big Cola bottle of hot water wrapped by the towel and cushion it on my neck. At first, I can not response quickly. I just say, "thank you very much", then the stewardess aunt kept asking me for many times if I need any other service and help me change one time of hot water! I can gradually feel the intimate care of the rating service from the airline.

其实设计也是一种服务，并不一定只是卖卖创意，好的体系，热诚的态度和执着的精神都会为你带来持续的回报。相信联合航空深谙此道，是值得尊重和学习的一个企业，让我牢牢地记住了她们的好。

Actually design is also one kind of service. We do not have to sell the creative point only. Good system and sincere attitude as well as persisting spirit, which will bring you lasting return. I believe that United Airline should know very clear about this. It is one enterprise which deserves everyone of us to respect and study. I will remember their kindness and goodness.

什么是经典与价值，不可超越的只有时间，经典与价值只能用时间去承载和传播，我们不妨多一些去消费这种“等级”服务，在享受的同时也想想怎么提升自己和企业的品质，那就值回票价了！

What is the classic and value? Only the time that can never be overtaken. Classic and value can only be inherited and disseminated by time. Why not let everyone of us take more time to consume and enjoy such kind of "Rating" Service? While enjoying the quality service, I think you will also think of issue like how to improve myself and the service quality of our enterprise. That is worthy of paying the ticket!

纽约华尔道夫酒店 *The Waldorf Astoria Hotel, New York*

小费文化，不可思议

美国，可以说是"小费"文化的代表大国。

真不习惯，开始也觉得很讨厌！也闹出不少小笑话。第一天在纽约的华尔道夫酒店吃午餐时，呼唤服务生，居然没有人理我。原来服务我们的那位侍者刚刚走开了，别的不服务我们，是小费的第一种。每个人小范围的"责任平安"的方式；第二天晚上去到了酒店的西餐厅吃大餐，刚好向一位年长的华人老侍者请教了解一下，也有两三个同时服务几台客人，小组式享受小费之方式，特别是较高档的用膳地方——唉，还有这么多的花样！

晚上再想想他们的结账方式和世相，吓了一跳。这种付费和小费其实是一种文明程度的体验，更可夸张地认为是诚信体系之最终表象。

落笔：特指吃饭，结账时，把钱放好在桌

账单上或本子夹着，你就可以走了，没有错，就这样。我不禁疑问：一、顾客会不会给少了，包括给了多少小费；二、钱会不会给别人拿走？三、会不会吃完没给钱或被完全没给钱的"霸王餐"……

小费，开始我不懂，其一是在美国不叫"Tips"，而叫"gratuity"，这让我拿起手机查了才懂；第二，它可以从你消费金额的最低5%到不设上限（导游也偏爱给我们讲小费奇遇的故事），一般档次越高，比例越高，当然也有让你自己填比例，或者账单上直接给你建议，你只需要选就好，18%左右，够可以了。当然，服务是你不可抗拒的！有时候会让你情不自禁地希望以这种方式来表达你的感激！

入住酒店后，把现金和证件锁在房间的保险箱里，三天后退房，开车近两小时到达西点军校，准备入内参观，突然发现居然忘记了这件事，让导游林先生打电话回酒店，心急如焚，路上美食无味，忐忑，这可是旅程的开始啊！！

待回到酒店，稍作等候，核对证件后，一叠从大到小数点好的钱和证件完完整整地交到我的手上（包括人民币和美元），无以为报，忙拿出一大张作小费，婉拒，再递出，仍没收下，他们说是他们份内事。原来小费的收与否还是有原则的。真不错，他们！开始留意关注这种行为的作用和得到的体验：体会到可以多给一些，服务体的人会更有价值感和认同感，相信对他（她）们的健康人生也有促进，原来这也可以说是美国大众教育的一种小方式！只有身处异地中才能深刻体会！以后在美国等小费国度消费可千万别忘了用小费表达你的情感和满意度，当然也是对这种习惯的尊重！

小费，真的不可思议，也令人佩服，背后的诚信体系更值得去研究！

Tip Culture, Incredible
小费文化，不可思议

纽约华尔道夫酒店 *The Waldorf Astoria, New York*

美国，可以说是“小费”文化的代表大国。

U.S. can be said the typical country which has the “tip” culture.

真不习惯，开始也觉得很讨厌！也闹不少小笑话，第一天在纽约华尔道夫酒店吃午餐时，呼唤服务生，居然没有人理我。原来服务我们的靓仔侍应刚刚走开了，别的不服务我们，这是小费服务的第一种方式——每个人小范围的“袋袋平安”的方式；第二天晚上专门订了酒店的西餐厅吃大餐，刚好向一位年长的华人老侍应了解一下，也有两三个同时服务几台客人，小组式享受小费的方式，特别是极高档用膳地方——唉，还有这么多的花样！

Actually I did not get used to it from the very beginning. I start to feel repugnant! Also I made quite a lot of embarrassment for this. The first day when I had lunch in the Waldorf Astoria Hotel in New York and I called the service man and no one care for me. The truth is that our service man had just left and other service men won't serve us. This is the first kind of tips. Every people knew the way of tips. The second evening, I specially ordered the western canteen for big meal, it came one senior Chinese man and I enquired from him and learned that there are 2 or 3 service men serve several guests in the same time, they got the tips in terms of group, especially the high-grade dining place — oh, My God. There are so many types of tips !

晚上再想想他们的结账方式和过程，吓了一跳，这种付费和小费其实是一种文明程度体验，更可夸张地认为是诚信体系的最简单表现。

At night, we just thought of the way of settlement and the process. I am so scared that this way of paying bill and tips actually is one experience of civilization. Over-exaggeratedly speaking, this is the most simple credit system.

消费完，特指吃饭，结账时，把钱放好在结账单或本子夹着，你就可以走了。没看错，就这样。我不禁疑问：一、顾客会不会给少了，包括给了多少小费；二、钱会不会给别人拿走？三、会不会吃完没给钱或者被冤枉没给钱吃“霸王餐”……

After finish the dinner and I need to pay the bill. Usually you can simply put the money in the bill or book clip, and then you can go. I can't help questioning. Firstly, didn't they worry about some customers may pay less? Can they know how much tips the customer gave? Secondly, didn't they worry about the money will be taken away by others? Thirdly, didn't they worry in case that some guests finished dinner and went away without paying the bill or some people may be treated unjusticly by eating without giving the money?

小费，开始看不懂，其一是在美国不叫“Tips”，而叫“gratuity”，这让我拿起手机查了才懂。第二，它可以从你消费额的最低5%，到不设上限（导游也偶尔向我们讲讲小费奇迹的故事）。一般档次越高，比例越高，当然也有让你自己填比例，或者账单上直接给你建议，华尔道夫嘛，18%左右，够厉害的。当然服务是你不可抵挡的！有时候会让你情不自禁地希望以这种方式来表达你的感激！

Tips, I really can not understand in the beginning. For the first reason, it is not called "tips" but called "gratuity" in U.S. I can only understand when I used my mobile to check this word. Secondly, it can be given from the lowest consumption 5% to non-limitation (the tourist guide also told us the story of tips). Usually the higher the grade, the proportion of tips will be higher. Of course some let your fill in the proportion by yourself, or it will directly give you the advice on the bill. For Waldorf, about 18%, of course service is quite excellent! Sometimes you can't help hoping that in this way to express your gratitude.

入住酒店后，把现金和证件锁在房间的保险箱里，三天后退了房，开车近两小时到达西点军校，准备入内参观，突然发现居然忘了这件事，让导游林先生打电话回酒店，心急如焚。路上美景无味，忐忑，这可是旅程的开始啊！！

After living in the hotel, I will lock the cash and certificate into the secure box. 3 days later, I checked out and took me nearly two hours' ride to arrive in the West Point Military School and prepared to visit inside. Suddenly I found that I forgot this issue. I asked the tourist guide Mr. lin to give a call back to the hotel. I was so worried. During the road, although outside window are the beautiful scenery, I felt unhappy, however, this is the start of the Journey!

VANDERBILT ROOM

纽约华尔道夫酒店 *The Waldorf Astoria, New York*

待回到酒店，稍作等候，核对证件后，一叠从大到小数点过的钱和证件完完整整地交到我手上（包括人民币和美元），无以为报，忙拿出一大张作小费，婉拒。再三表示，亦没收下，他们说是他们分内事，原来小费的收与否还是有原则的，真小看了他们！开始沿途关注这种行为的作用和得到的体验：你高兴了可以多给一些，服务你的人会更有价值感和认同感，相信对他（她）们的健康人生也有促进，原来这也可以说是美国大众教育的一种小方式！只有你身处其中才能深有体会！以后在美国等小费国度消费可千万别忘了用小费表达你的情感和满意度，当然也是对这种习惯的尊重！

When coming back to the hotel, please just wait. After finishing checking the certificate, from big stacks of money to small stack of money and certificate, they gave me all these completely (including RMB and U.S. Dollar). I can not know how to return this goodness. I took one big piece as the tips and being rejected. I gave him once again and they did not get it. They told me that's the basic job they should do. I came to know that charging the tips or not also have the rule. Don't look down upon them! I start concerning this effect of this behavior and got the experience. If you are happy, you can give more, and the one serviced you will feel more sense of value as that means you recognize their service. I believe that it may help promote their healthy life. I never know that this also an American way of mass education! Only if you are inside can you have the deeper understanding! Later on, if you travel in U.S. later, please don't forget to give the tips to express your feeling and your satisfaction. Of course this also shows your respect to this behavior!

小费，真的不可思议。也令人佩服，背后的诚信体系更值得去研究！

Tips, it is just incredible. I admire this and the credit system behind this also should be worthy of studying further!

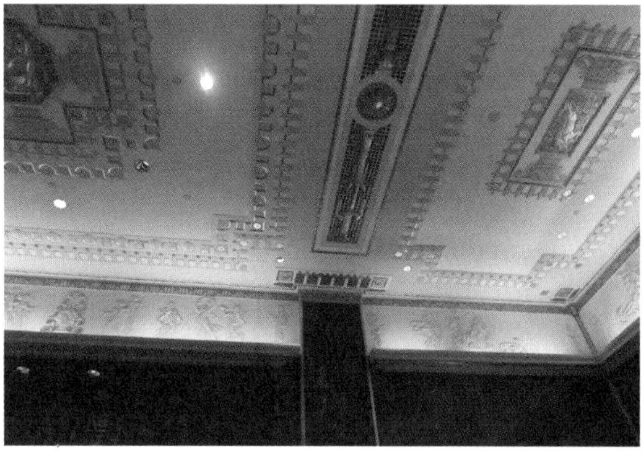

纽约华尔道夫酒店 *The Waldorf Astoria, New York*

BOSTON HARBOR HOTEL
AT ROWES WHARF
波士顿港湾酒店
★★★★★

7

BOSTON HARBOR HOTEL, BOSTON

Address : ROWES WHARF ,70 ROWES WHARF, BOSTON, MA 02110
Telephone : 617-439-7000
Http : //www.bhh.com/

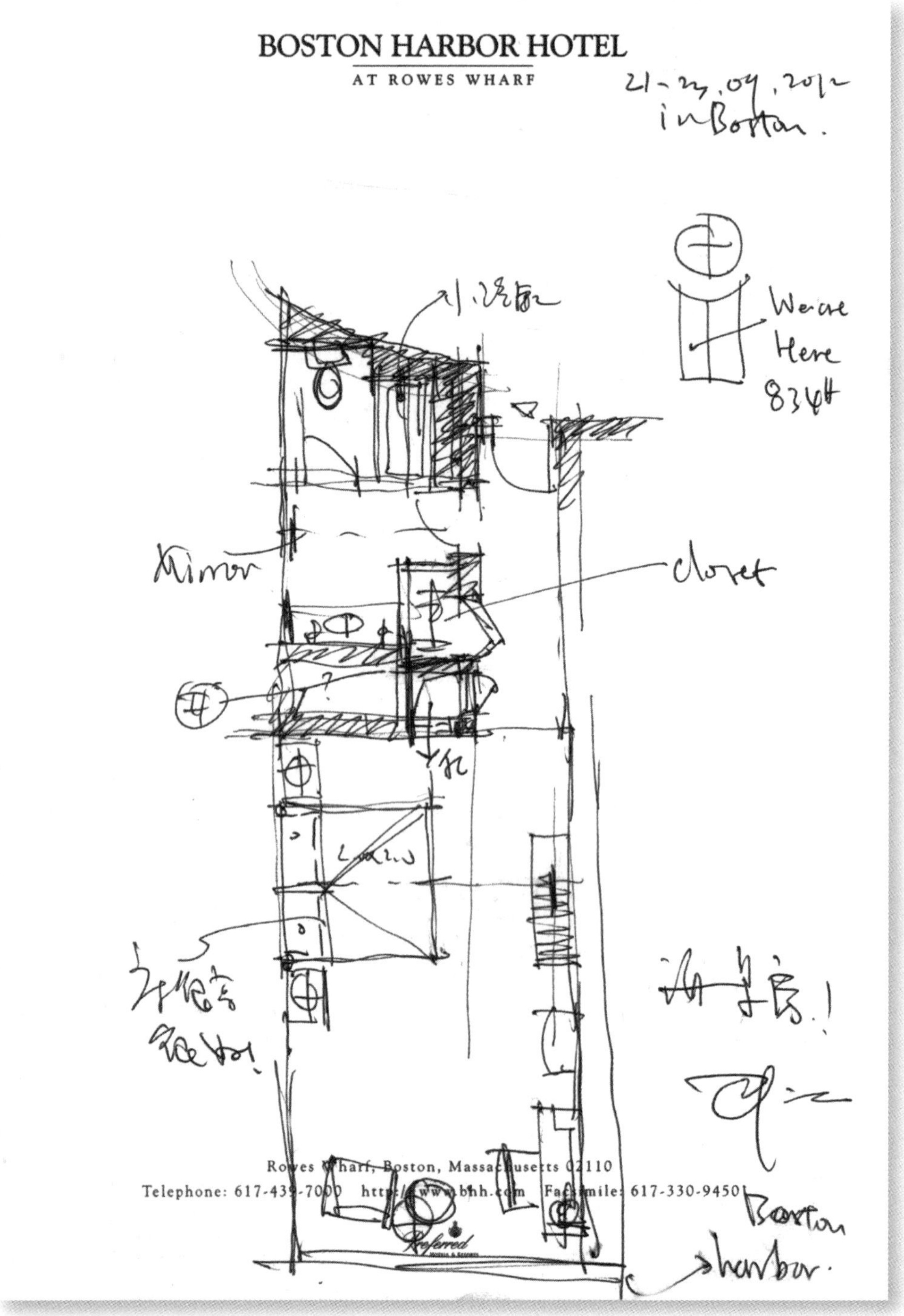
BOSTON HARBOR HOTEL
AT ROWES WHARF
21-23, 09, 2012
in Boston.
We are Here
834#
Mirror
closet
Boston harbor.
Rowes Wharf, Boston, Massachusetts 02110
Telephone: 617-439-7000 http://www.bhh.com Facsimile: 617-330-9450

怎样的城市最宜居

什么样的才算是宜居的城市？有好的天气：蓝天白云，温度适中，湿度宜人；有好的风景：沙滩、绿化、雅致的建筑、蜿蜒的街道；有好的食材：海鲜、农产品，当然最关键是有人文特色：好的历史背景和友善的当地居民，哈哈，差一点漏了，世界级的名牌大学，波士顿就有：哈佛、麻省理工，当然也就更是闻名于世。

从New York高速公路上大约开了5个小时才到达波士顿，近傍晚，给人以舒心的第一印象。第二天市区作步观光，看到各种小景点、各种博物馆，也有人气的大市场，有自行车比赛（可惜之后不久的马拉松比赛发生爆炸事件，向遇难的人致哀），有古典味十足的酒店，借用洗手间时顺便参观了一下，什么品牌，有些小物件，市政府、教堂，等等，值得逛的地方三天逛一个，在不经意中让你的眼睛得到享受！

专门挑选面海的房间，可以晚上对着大海发

果，咖啡，水手，轮船，这样悠游在你的节奏，不然怎能体味透过玻璃的海水粼粼之光呢？！

波士顿就是一个宜居的滨海城市！

TRASH
ONLY

What is the Most Livable City?
怎样的城市最宜居

NORTH MARKET

什么样的才算是宜居的城市?有好的天气：蓝天白云，温度适中，湿度宜人；有好的风景：滨海、绿化、细致的尺度、蜿蜒的街景；有好的食材：海鲜、农产品；当然最关键是有人文特色：好的历史背景和友善的现住居民，哈哈，差一点漏了。世界级的品牌大学波士顿就有：哈佛、麻省理工，当然龙虾更是闻名于世。

What makes the most livable city? Good weather, blue sky and white clouds, moderate temperature, humidity and pleasant; good scenery: the coastal, greening, detailed scale, winding streets; good food available like seafood, agricultural products; the most important thing is the humanistic characteristics: good historical background and friendly and amicable permanent residents, haha, narrowly missed this point. The world-class branded University Boston has all these, Harvard, MIT, of course, lobster there is quite well-known to the world.

OMNI HOTELS
THE PARKER HOUSE

从New York高速公路上走走堵堵近5个小时才到达波士顿，近傍晚，给人以舒心的第一印象。第二天市区徒步观光，看到各种小景点，各种小商铺，也有人气的大市场，有自行车比赛（可惜之后不久的马拉松比赛遭受炸弹袭击，向逝去的人致哀）。有古典味十足的酒店，借用洗手间时顺便参观了一下，什么品牌，有照片为证，市政厅、教堂等等，值得驻足的地方三五步一个，在不经意中让你的眼睛得到享受！

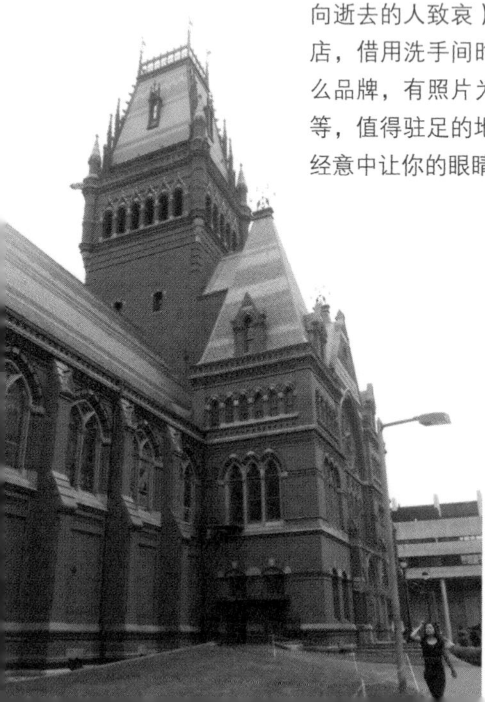

From the New York highway and turn around the walls of traffic jam and it takes me nearly 5 hours to arrive in Boston. It's near the evening; this city gave me the impression of relaxing and comfortable. On the second day, I walked on foot sightseeing the city downtown and saw all kinds of scenic spots, all kinds of small shopping booths, also the popular big market, a bicycle race (unfortunately, shortly after the marathon suffered bomb attacks, people show the sorrow in order to honor the dead). Hotel with classical flavor, I dropped in on my way of going to the washing room. All kinds of brands, as shown in the picture, the city town hall, the church, and so on, the very good place for visiting can be everywhere, you can easily see it every three or five steps, just let your eyes enjoy the beautiful scenery unconsciously!

波士顿港湾酒店 *Boston Harbor Hotel*

专门挑选面海的房间，可以日夜对着大海发发呆，咖啡、水果、杂志，这才是旅游应有的节奏，不然怎能体味透过玻璃看看海水粼粼之光呢？！

We select specially the room facing the sea, you can face to the sea day and night in a daze.Coffee, fruit, magazine, this is the due rhythm for tourist, otherwise how can you appreciate the sparkling sea water through the glass ?

波士顿确实是一个宜居的滨海城市！
Boston is indeed a livable coastal city!

SENECA NIAGARA

CASINO & HOTEL

塞内卡尼亚加拉酒店

★★★

8

SENECA NIAGARA CASINO & HOTEL

Address : No.310 fourth street niagara falls, ny 14303
Telephone : +716-299-1100
Fax : +716-278-3699
Http : //www.senecaniagaracasino.com/

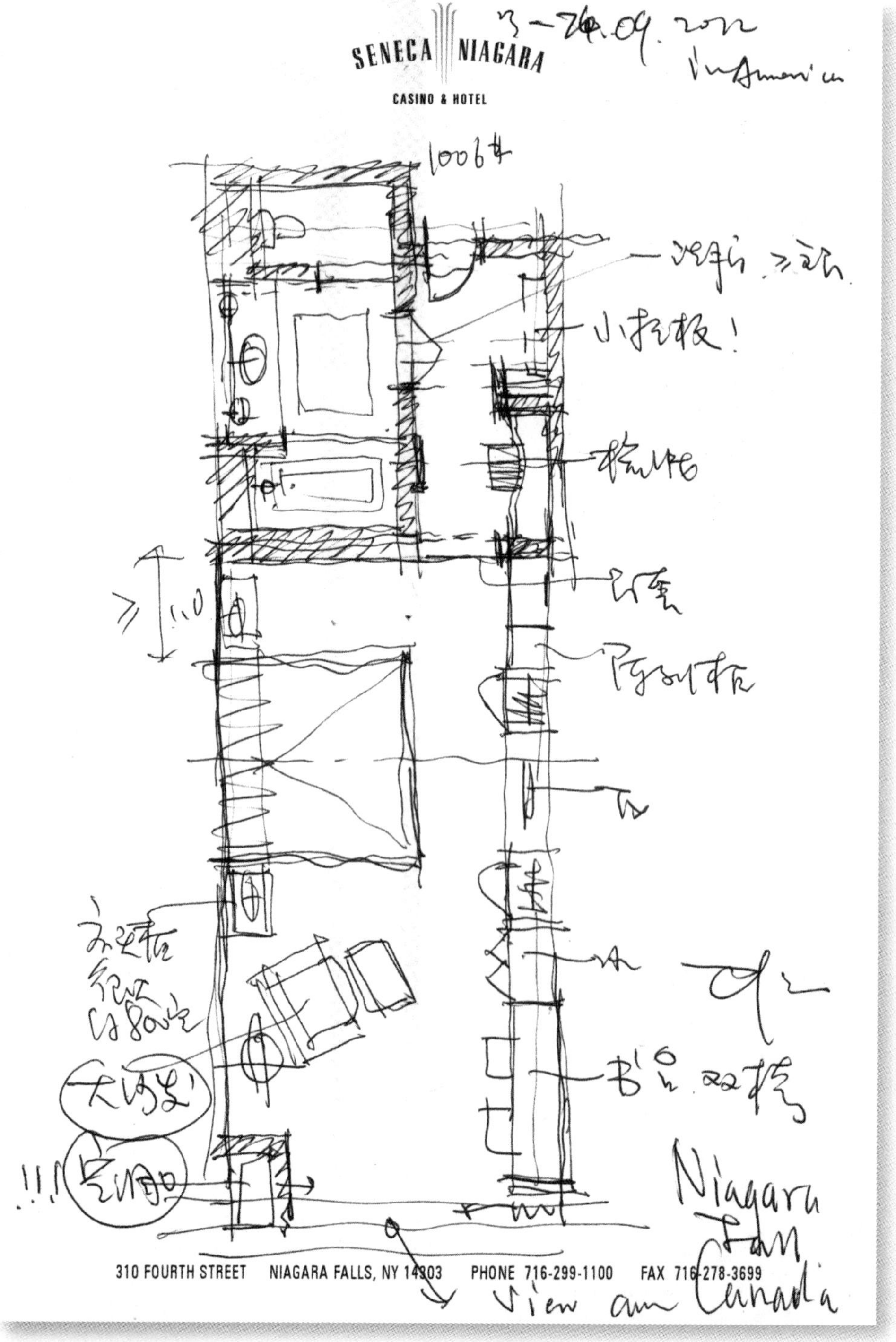
SENECA NIAGARA
CASINO & HOTEL
1006#
Niagara
310 FOURTH STREET NIAGARA FALLS, NY 14303 PHONE 716-299-1100 FAX 716-278-3699

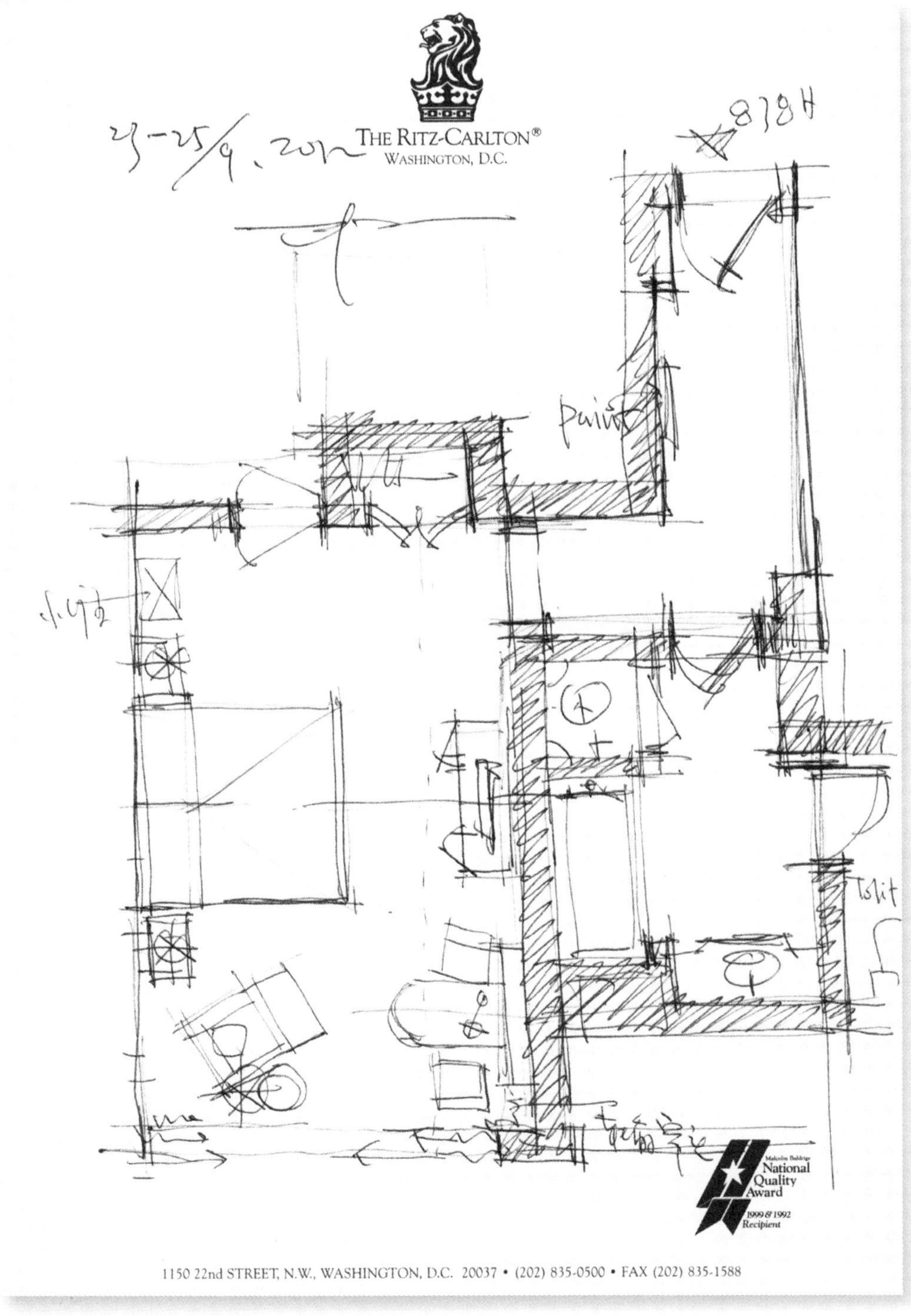
23-25/9, 2012
THE RITZ-CARLTON®
WASHINGTON, D.C.
8784
Toilit
Malcolm Baldrige National Quality Award
1999 & 1992 Recipient
1150 22nd STREET, N.W., WASHINGTON, D.C. 20037 • (202) 835-0500 • FAX (202) 835-1588

THE RITZ-CARLTON®

华盛顿丽思卡尔顿酒店

★★★★★

THE RITZ-CARLTON HOTEL, WASHINGTON.D.C

Address : No.1150 22nd STREET, N.W.,WASHINGTON,D.C. 20037

Telephone : (020) 835-0500

Fax : (020) 835-1588

Http : //www.ritzcarlton.com/

在年轻的国度看历史

去华盛顿，看的是各式各样的建筑物，纪念碑、博物馆、图书馆、艺术区，但客观地说，美国只是一个年轻的国度，但其留存下来的好像也不少比我们古老的！是不是应该反省一下我们留下古建筑太少了。不能都是在大山大岭里面的庙、堂、祠、窟之类的，要像美国一样可以供人参观流览，真正的才好！

选择了古朴的酒店，2000年开业的华盛顿丽思卡尔顿酒店，内敛的北美风格，可说大本色。在一条叫中某街的"酒店街"上，入住后悠闲地在宾馆比邻林立的街道上溜达，看看小店铺、服装店、家具店、画廊，找找临近做旧的古本行，看看酒店：Westin、Fairmont、Park Hyatt等等，饱了口福也饱了眼福！现代的、古典的、艺术的各有不同：

酒店的房间在走道的尽头，面向自家的后花园，非常宁静，服务生也礼貌得很，房间的布局设计是我喜欢的，也很人性化，没有门的衣帽间睡眠区，声光的干扰降到最低，时差、半夜打电话回国的安排的工作也不会影响到同住的人，确是上佳的布局方式！

可惜游泳池在维修，早上没有到泳池，但是安心地享受了一个美味的美式健康早餐，也顺便读了一份晨英语，看来年纪小的也不会挨饿了，因为是单点式的早餐，美国的酒店多用这种方式，希望我们的酒店也可以学习一下，之前在欧洲的已经看到，欧洲的酒店都有这种方式，相对于我们国内酒店的开放式自助餐，这种更节约、更减少浪费。

从细节中看全面，有时候优良的文化和管理的传承也是形成历史的一部分，在这个瞬息万变的国度中，学习历史，不可忘，更是一种自信，对我们更是一种提醒。

提醒我们将优秀的历史传承，用什么载体，以什么方式！？

华盛顿丽思卡尔顿酒店 *The Ritz-Carlton, Washington D.C.*

Look at the History in the Young Country
在年轻的国度看历史

去华盛顿，重点是看看一些建筑物：纪念碑，博物馆，图书馆，老街道，但客观来说，美国只是一个年轻的国度，但其留存下来的好像有不少让我们去学习的！是不是应该反省一下我们留下的古建筑有多少，不能都是在大山大岭里面的庙、堂、祠、窟之类的，要像美国一样是可以学习交流的，亲近的才好！

Go to Washington, of course, we need to see some buildings such as monuments, museums, libraries, old streets. Objectively speaking, America is just a young country, but its remaining seems to have many to let us to learn! Is it right? Didn't we reflect how many ancient building remained for us? They are temple, church, ancestral temple, cave like in the mountain, that's not the way we see the civilization. It should be communicated and close as that of America.

选择了老牌的酒店，2000年开业的华盛顿丽思卡尔顿酒店，内敛的北美风格，低沉的木色，在一条旺中带静的“酒店街”上。入住后悠闲地在商业味浓郁的街道上漫步，看看小店铺，服装店、家具店、画廊，找找吃的，顺便做做老本行，看看酒店：Westin、Fairmont、Park Haytt等等，饱了肚福也饱了眼福！现代的，古典的，艺术的各个不同。

Chosen the old brand of hotel, The Ritz-Carlton in Washington D.C, which opened in 2000,bear introverted North American style,deep wooden color,and is located. at a busy but quiet "hotel street", You can stroll on the street with thick commercial flavor leisurely after living there, Looking at the small shops, clothing stores, furniture stores, art galleries,.looking for something to eat ,and backing to the old prefession passingly by looking at the hotels: Westin, Fairmont, Park Haytt, and so on, thus had a nice feast of stomach and eye! Modern, classical ,art, quite different.

酒店的房间在走道的尽端，面向自家的后花园，非常宁静，服务生也礼貌得很。迂回的布局方式是我喜欢的，也很人性化。洗手间远离睡眠区，声、光的干扰降到最低。时差，半夜打电话回国内，安排沟通工作，不会影响到同住的人，确是上佳的布局方式！

My hotel room is at the end of the corridor, facing to their rare garden. It's peaceful and quiet. The bell boy is quite polite. The roundabout layout is my favorite. And it is humanized, too. The washing room is far away from sleeping area, so that disturb of noise, light from washing room can be limited to be smallest. With time different, I must make Chinese call in the middle of the night to communicate about work arrangement, but I will not affect the person in the same room. This layout is really perfect!

可惜游泳池维修，早上没有了晨练，倒是安心地享受了一个美味的美式健康早餐，也顺便练了一个日常英语，看菜单的水平也小有提高了。因为是单点式的早餐，美国的酒店多用这种方式，看来我们的酒店也可以学习一下。之前在休闲的巴厘岛，欧洲的酒店都有这种方式。相对于我们国内盛行的开放式自助餐，这种更新鲜，更减少浪费。

It's a pity that the swimming pool was under repair. I can't do exercise in the morning, but I was reassuring to enjoy a delicious American breakfast, and practice my daily English incidentally. I also improved my level of reading the menu. Breakfast is ordered from menu. Most American Hotels use this way for breakfast. I think our hotels can learn about it. And some hotels in the leisure Bali Island and in European are also in this way. Comparing to the open buffet of breakfast which is popular in China, this way is more fresh and good for waste reduction.

从细节中看全面，有时候优良的文化和简单的传承也是形成历史的一部分，在这个先进的、年轻的国度中学习历史，不可笑，更是一种途径，对我们更是一种提醒。

Fully from the detail part, sometimes excellent culture and the simple inheritance is part of formation of history. Learning history from this young developed country is not ridiculous. It is also a way of learning and sort of reminder for us.

提醒我们将优秀的品质传承，用什么载体，以什么方式？！

It can remind us to inherit the excellent quality, and make us to think the carrier and way for inheritance?

MANDARIN ORIENTAL
LAS VEGAS
拉斯维加斯文华东方酒店
★★★★★

10

MANDARIN ORIENTAL, LAS VEGAS

Address : No.3752 Las Vegas Boulevard South, Las Vegas, NV 89158 USA+1
Telephone : + (702) 590 8888
Fax : + (702) 590 8880
Http : //www.www.mandarinoriental.com/lasvegas/

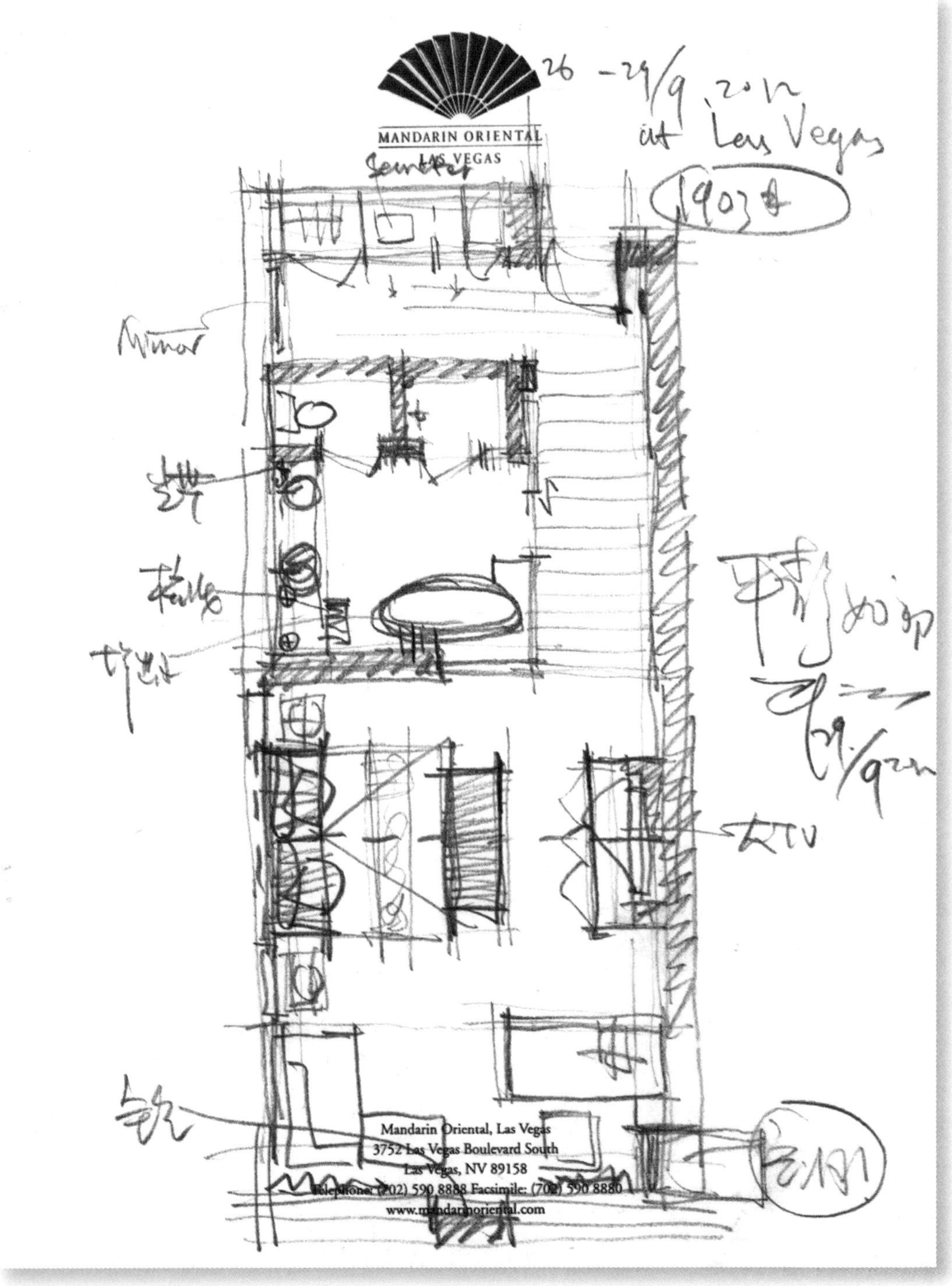
MANDARIN ORIENTAL
LAS VEGAS
26-29/9, 2012
at Las Vegas
1903
Mirror
KTV
Mandarin Oriental, Las Vegas
3752 Las Vegas Boulevard South
Las Vegas, NV 89158
Telephone: (702) 590 8888 Facsimile: (702) 590 8880
www.mandarinoriental.com

精彩如初

"因为喜欢，所以欢喜！"

我连续在两个城市选住了位于中心区的Mandrin Oriental文华东方酒店，先是赌城Las Vegas，再到同样的三藩市的，酒店保持有东方的韵味，特别是开张不多久的拉斯维加斯的这一间，从入口前厅的门厅大堂，再到空中的接待大堂，套房、水疗功能区再到走廊、客房……

推门而入，更惊艳于淡雅的色调，让人宁静，乍一看去也不会让你惊喜，但下来就不同了：窗外虽是闹市繁华区，但你听不到一丝噪声，柔和的背景音乐让你回家般的放松，它的spa更是数一数二，可惜没有安排时间去享受，典型的大衣帽间，随意可摆放大大的旅行箱，温馨调和的洗手间，细节如家，低调的香薰，依窗而舒畅，真正让你体验"精彩如初"的感觉！

每次出国旅行可以保留住这个品牌，更多的时候，能让你住到正宗中式的西方的文华东方酒店！

Comfortable as Usual
平静如初

“因为喜欢，所以欢喜！”

“We are happy just because of love!”

就连续在两个城市选住了位于中心区的Mandrin Oriental文华东方酒店，先是赌城Las Vegas，再到后来的三藩市的。酒店保持有东方的韵味，特别是开张不多久拉斯维加斯的这一间，从入口首层的门厅大堂，再到空中的接待大堂、餐厅、配套功能区，再到走廊、客房……

Choose Mandarin Oriental Hotel in the city downtown successively in 2 different cities. First of all, it was the Casino City in Las Vegas, then came to San Francisco City. The hotel still keeps the oriental taste, especially the Las Vegas, which is not opened for a long time, from the entrance of the lobby at the first floor, then the Reception Lobby in the air, canteen and any other functional area, then the corridor, guestroom…

推门而入，更是延续深褐色调子，让人宁静，乍一看一点也不会让你惊喜，住下来就不同了。窗外虽是纷闹的商业区，但几乎听不到一丝噪声，柔美的背景音乐让你回家般的放松（它的Spa更是数一数二，可惜没有安排时间享受）。典型的大衣帽间，随意可摆放大大的旅行箱。温馨调子的洗手间，细节如家，微微的香薰，依窗而斜躺，真正让你体验“平静如初”的感觉!

拉斯维加斯文华东方酒店 *Mandarin Oriental , Las Vegas*

Pushing the door and going inside, it continues the dark brown tone, which keeps people calm. Just taking a glance, you won't feel any surprise, but living here is quite different. Although the view through the window is busy commercial district, however, almost we can not hear the noise. The soft underground music will make you relaxing just like coming back home (Spa is quite good and I think it is second to none. What a pity that we can not take more time to enjoy it). Classic room for big clothes room and bigger luggage can be still placed there randomly. Washing room is cozy and comfortable and the considerate details make you feel like staying at home. Slight fragrance, leaning by the window and lying on the bed, which let you experience that kind of feeling sense of "comfortable as usual".

每次出国旅行可以保留住这个品牌，思乡的时候，能让你吃到正宗中式的面条的文华东方酒店！

Every time you need to travel abroad, you can keep this brand. Whenever you feel homesick, you can eat the authentic Chinese noodle in Mandarin Oriental Hotel!

The GRAND AMERICA Hotel

拉斯维加斯盐湖城酒店

★★★★★

11

GRAND AMERICA HOTEL, LAS VEGAS

Address : No.555 South Main Street
Salt Lake City, Utah 84111

Telephone : 801-258-6000

Fax : 801-258-6911

Http : //www.grandamerica.com/

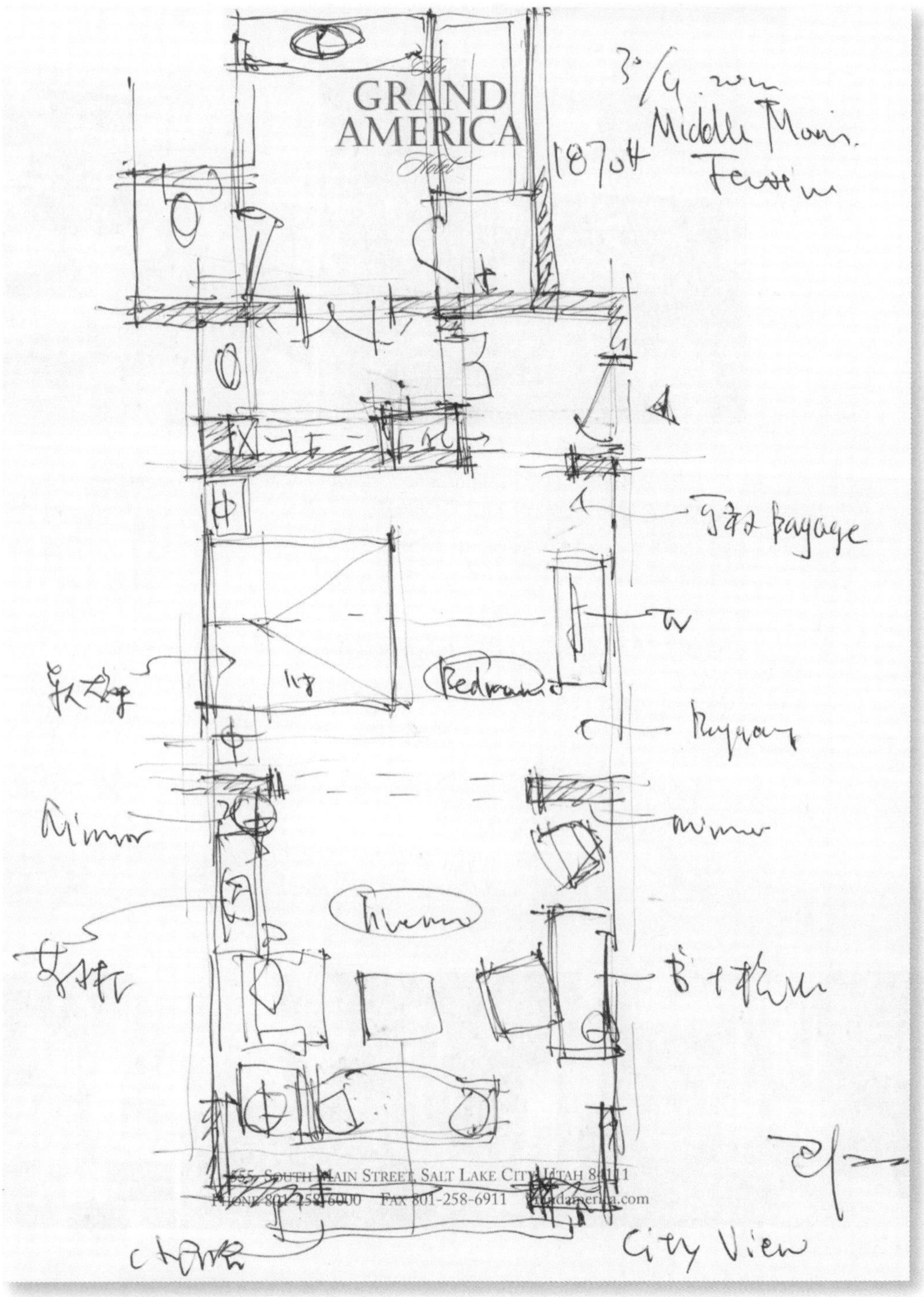

City View

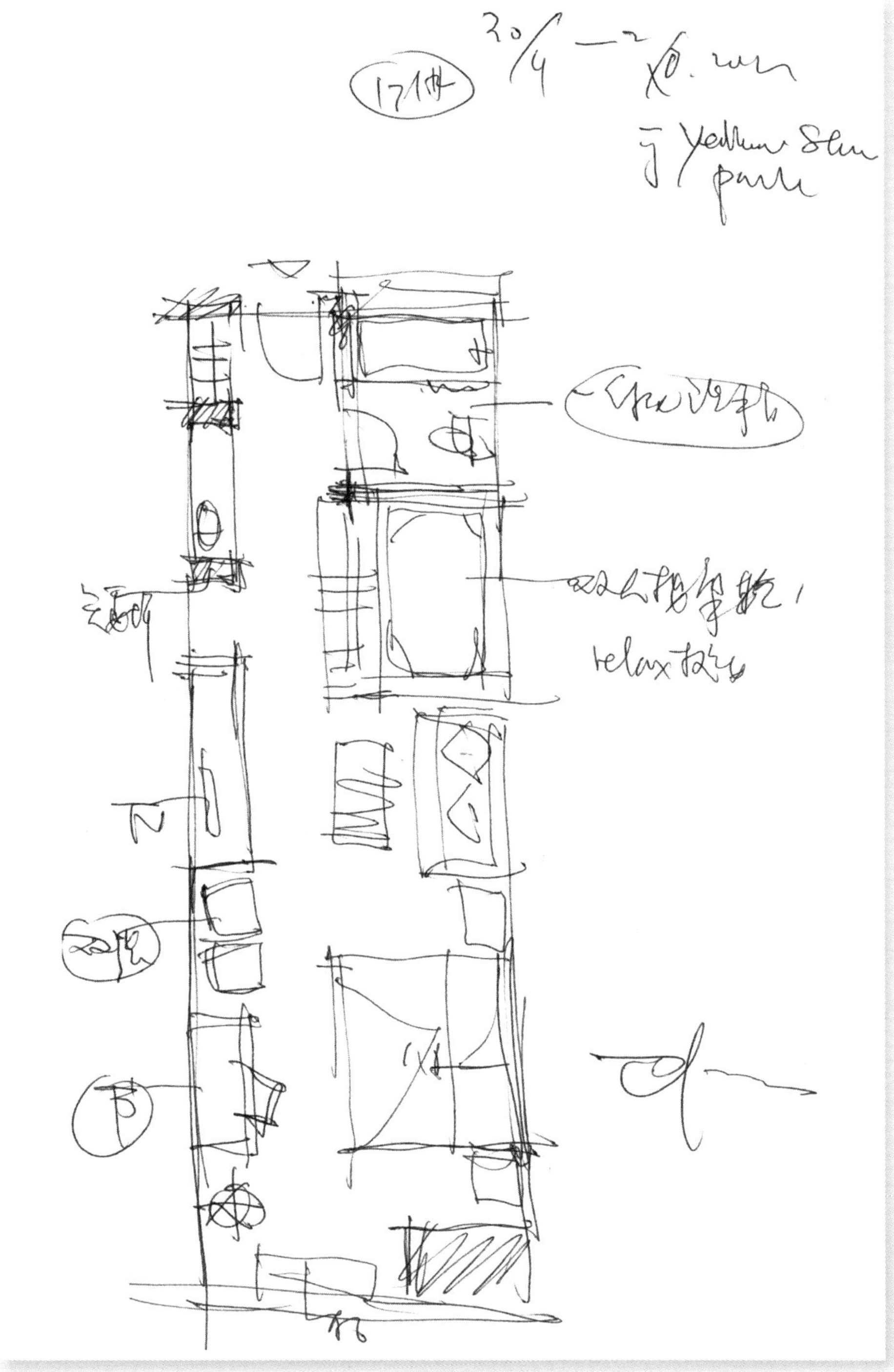

Holiday Inn

黄石公园假日酒店

12

★★★

HOLIDAY INN, WEST YELLOWSTONE

Address : No.315 yellowstoe Avenue
West Yellowstone,
MT 59758

Telephone : +1 406-646-7365

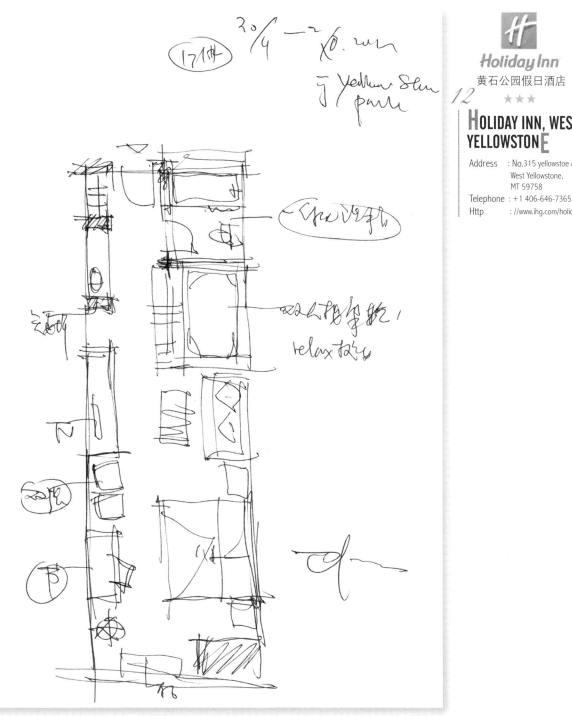

Http : //www.ihg.com/holidayinn/hotels/

有趣的浴缸

我在黄石公园没找到更好的酒店，我将就着在西黄石的小镇上入住这家假日酒店，实用经济式连锁酒店，小小的简单的大堂，小小的便利店，精致的顾客餐厅，长长的走廊，当然也有自助的洗衣区、冰粒机等，足以满足日常需求，也适合长居使用的设施。

倒是进入房间还是有不少的不同，宽宽长长的，布局也有特色，特别适合结伴旅行的人士：设在走道的洗手盆方便入房即洗脸、去垢、去泥；独立的淋浴可以满足两人同时分别使用，最绝的是放在起居区侧的大大的双人按摩浴缸，相当"震撼"，与廉价收费的酒店格格不入，很酷了！

晚上大大地享受了一次热水按摩，中央四台，仅有的中文电视台陪伴，一天的疲劳尽消，可谓是考察间隙，也算了在假日酒店的惊喜收获吧！

Interesting Bath Tub

有趣的浴缸

黄石公园假日酒店 *Holiday Inn ,West Yellowstone*

就近黄石公园没找到更好的酒店，就将就着在西黄石的小镇上入住了这家假日酒店，实用经济式连锁酒店，小小而亲和的大堂，小小的便利店，雅致的配套餐厅，长长的走廊，当然也有自助的泳池区，冰粒机等既满足日常需求，也宜长居使用的设施。

I can not find one better hotel near the Yellowstone National Park. So I lived in this Holiday Inn in this West Yellowstone small town, which is one economic type of chain store hotel. Small but amicable lobby, small convenient store, elegant canteen with full equipment, long corridor, of course, there is self-service swimming pool, ice machine to meet the daily requirement, also it is suitable for the people who needs to live for a longer time.

倒是进入房间还是有不少的不同，窄窄长长的，布局也有特色，特别适合徒步旅行的人士：设在走道的洗手盘方便入房即洗脸、去尘、去泥；独立洗浴可以满足两人同时分别使用，最绝还是放在起居室区侧的大大的双人按摩浴缸，相当“震撼”，与这种收费的酒店“格格不入”，很吸引人！

There are so many differences when entering the room, narrower and longer. The layout has its own characteristics, especially suitable for the people who love walking or hiking. The sink that equipped in the aisle is for the convenience of getting into the room, i.e. washing, de-dusting, de-earth. The independent shower can satisfy two people to use in the same time separately. The most wonderful thing is the adult two-people massage bathing tub at the side of the living room, which is quite sense of "shock", this is totally out of expect and quite attractive to the customers!

黄石公园假日酒店 *Holiday Inn ,West Yellowstone*

当然是大大地享受了一次热水按摩，中央四台，仅有的中文电视台陪伴，一天的疲劳尽清，确是考虑周到，也算在假日酒店的惊喜收获吧！

Of course, the most enjoyable thing is enjoying the hot water massage. With the company of CCTV4 , wich is the only Chinese Channel available in the hotel, the fatigue and tiresome during the day are entirely gone. It is indeed considerate and thoughtful. It gave me so many surprises when living in the Holiday Inn!

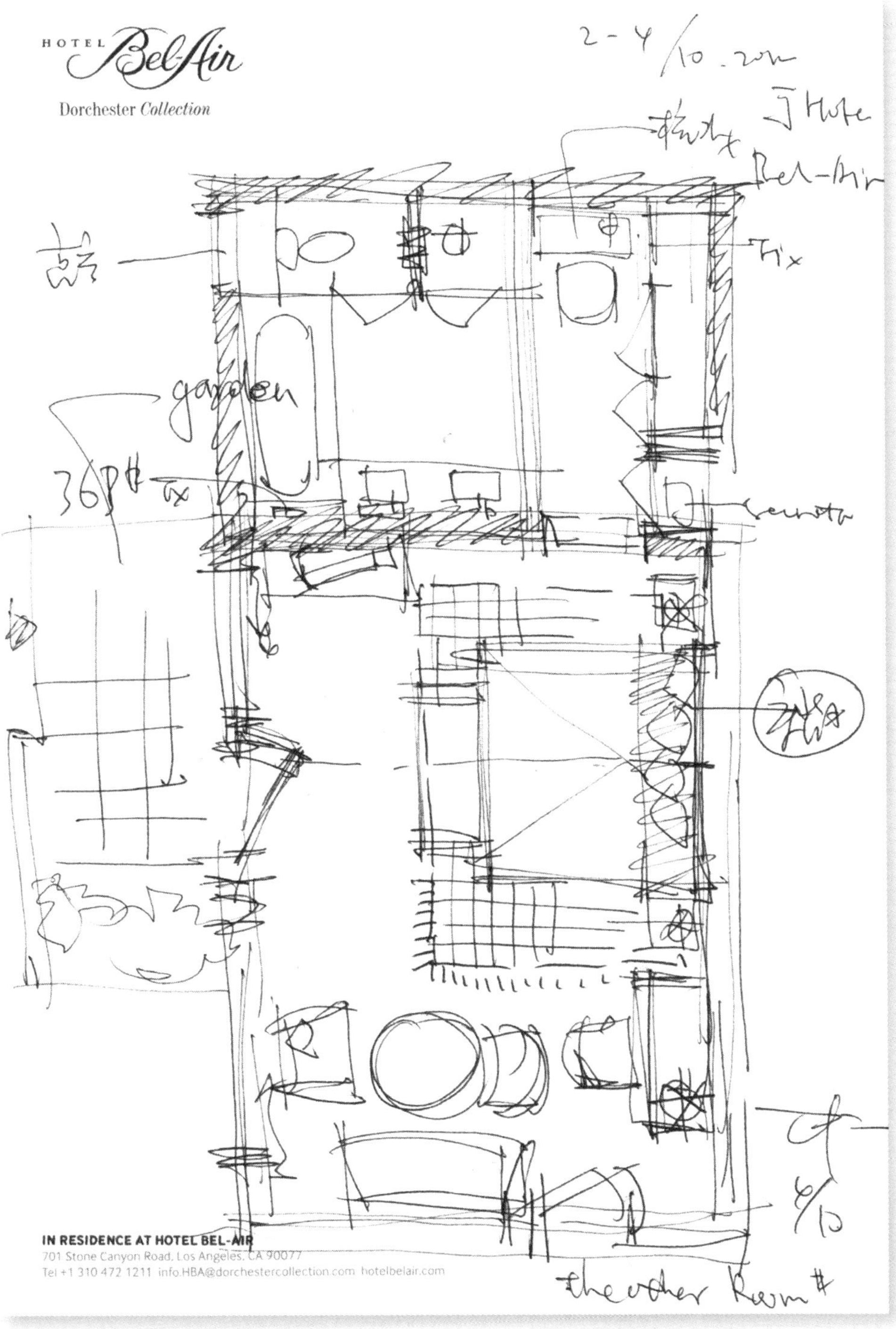
HOTEL Bel-Air
Dorchester Collection
garden
the other Room #
IN RESIDENCE AT HOTEL BEL-AIR
701 Stone Canyon Road, Los Angeles, CA 90077
Tel +1 310 472 1211 info.HBA@dorchestercollection.com hotelbelair.com

HOTEL Bel-Air
Dorchester Collection

洛杉矶贝尔艾尔酒店

13 ★★★★

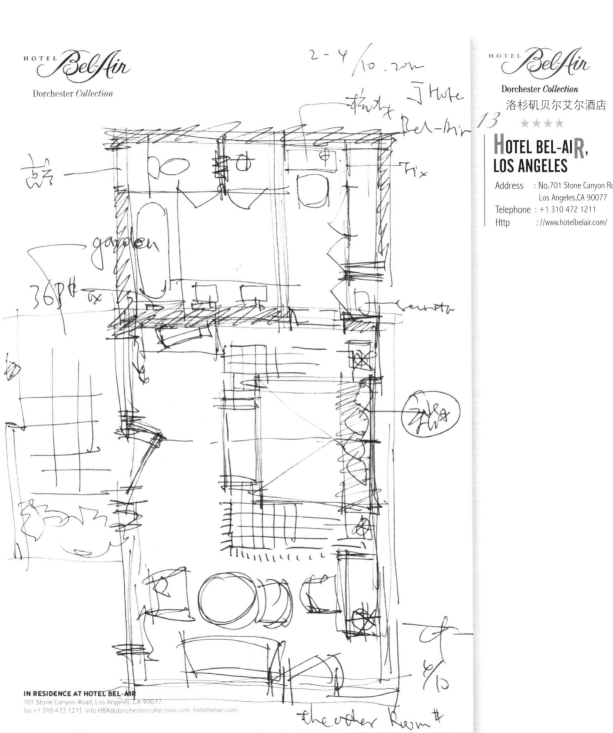

HOTEL BEL-AIR, LOS ANGELES

Address : No.701 Stone Canyon Road, Los Angeles,CA 90077

Telephone : +1 310 472 1211

Http : //www.hotelbelair.com/

难忘绿树径的粉色小屋

找酒店，通过酒店达人网先看看评论，人气高的排前面；再看看价格，贵的排前面；更看看地点，名人聚集的地方，再看看图片，绿树丛中的西班牙式粉色屋，就选了它——比华利山区域，日落大道的这间酒店，Hotel Bel-Air中文叫贝莱尔酒店。

说来有趣，近日再次翻看日本建筑师浦一也先生的《旅行从客房开始》，原来他列为第一个案例，被这位"变态"设计师列为世界第一的酒店就是这一家，真是孤陋寡闻了。但细看他是改造前的1996年住的，相隔16年之后而我们住的是在它被Dorchester集团收购后，请大师重新设计更新后的作品，怪不得我后来不知道它已经有65年的历史，失敬失敬！但从入口的名车林立就知道这是个高贵的"世外桃园"，导航也找了很久才找到进去的私家林荫小道，相信我们是少有自驾车来入住的客人，看来做足旅游前的功课还是很重要的，不然就错过了玛丽莲·梦露也曾住过的这间贝莱家酒店了！

入住的客房是个套房，前面独立不说的私家小花园，紧靠中心花园，如沐一世光生所写的，大树上真的有好些松鼠、大鸟、小鸟，房间大大面的窗，床正正对着花园，当然还是下着窗帘感觉安全私隐一些，高大舒适的设计，非常不起眼，但很有细节，更重要是用钱也买不来的套件：Bose的平板电视，ipad控制全屋设备，kelista（科勒的高档品牌）的洁具齐全，有设计品位的台灯，舒适经典的名品家具，……室内设计师 Alexandra Champalimaud 真是业内高手，不经意地打造了极致的空间。

其实让我记住的倒不是这些，因为它们可确实极其和谐地融在一起，不显眼，只是眼光刁馋才会关注到。次日到处走走，从山下到山上看了它各种房间，还有Spa（专业，可惜未及去享受）绿树成荫，成林，让这些粉色西班牙建筑散落在其中，静中活泼，流连忘返，当然，在这三天，尽情享受它的阳光，花园、树园泳池（一个人游泳），晚上也体验大厨

Wolfgang Duck打造的全套餐，太精彩了。一菜式配一种酒，“酒不醉人，人自醉！”深深体味了(三文治)“什么是钱”生活，撑死了，一顿晚餐吃了四小时！

粉色，性感而青春，让人难忘！

Memorable Pink House in the Green Tree

难忘绿树里的粉色小屋

洛杉矶贝莱尔酒店 *Hotel Bel-Air, Los Angeles*

找酒店通过“酒店达人网”，先看看评论，人气高的排前面；再看看价格，贵的排前面；更看看地点，名人聚集的地方；再看看图片，绿树丛中的西班牙式粉色屋，就选了它——比华利山区域，毗邻日落大道的这间酒店，Hotel Bel-Air中文叫贝莱尔酒店。

Searching the hotel via the "Hotel Master Net", have a look at the comments firstly, front row is the popularity, then check the price, the expensive one came the first, then change the location, the place for the celebrity gathering, then have a look at the picture, the Spanish model of pink house in the green wood, then we choose it-Beverly Hills area, the hotel is adjacent to Sunset Avenue, Hotel Bel-Air in Chinese called Beller Hotel.

说来有趣，近日再次翻看日本建筑师浦一也老先生的《旅行从客房开始》，原来他列在第一个案例，被这位“变态”学者列为世界第一的酒店就是说这一家，真就是孤陋寡闻了。但细看他是改造前的1996年住的，相隔16年之后，而我们住的是在它被 Dorchester 集团收购后，请大师重新设计更新后的作品，怪不得我压根不知道它已经有65年历史，失敬失敬！但从门口的名车林立就知道这是个高奢的“世外桃源”。导游也找了很久才找到了进去的私家林荫小道。相信我们是少有乘商务车来入住的客人。看来做做旅游前的功课还是很重要的，不然就错过了玛丽莲·梦露也曾住过的这间贝莱尔酒店了！

The most interesting thing is that recently I red once again the book with name "Journey starts from the guestroom" codified by Japanese Architect Mr. Puyiye. This hotel is listed as the first case in his book. Such kind of hotel is classified as the world No.1 hotel by this "metamorphosis" scholar. I felt that my knowledge is quite limited with scanty information. However, when we studied it carefully that he lived in 1996 when it was the time before the transformation. 16 years later, after it was acquired by Dorchester Group and they had this hotel renovated and this is the works after being renovated and updated. No wonder I didn't know that it already had a history of 65 years. So sorry for that! However, from the parking lot where is parked with so many branded cars and we get to know that it is also one luxurious "Quiet Land". It took the tourist guide quite a long time to find the private shade lane. I believed that seldom guests will be like us taking a commercial vehicle to come here. So it is quite important to make full preparation first before you travel. Otherwise you will miss Hotel Bel-Air, where ex movie star "Marilyn Monroe" had ever lived here as well.

入住的客房只有一层，前面微微下沉的私家小花园，紧接中心花园，如浦一也先生所写的，大树上真的有好些松鼠、大鸟、小鸟，房间大大面的窗，床正正对着花园，当然还是下着窗纱感觉安全私隐一些，高大简洁的设计，非常不起眼，但很有细节，更重要的是用钱堆出来的奢华：B&O的平板电视，IPAD控制全屋设备，Kallista（科勒的高端品牌）的洁具五金，有设计品位的台灯，舒适经典的名品家具……室内设计师Alexandra Champalimand真是业内高手，不经意地打造了极致的空间。

There is only one storey for guestroom. The slightly sinking area in the front is the private garden, close tightly to the center garden. As Mr. Puyiye described, there were some squirrels, big birds, small birds, room with bigger window, bed is just opposite the garden, of course the better to cover with window chiffon for more secure and private, higher and concise design, quite simple but detailed. The most important thing is that it was the luxury stacked by the money: B&O flat TV, IPAD equipment for controlling the entire house, Kallista (high-end brand of Kohler) sanitary hardware, table lamp with designing taste, comfortable and classic famous furniture…Interior decoration designer Alexandra Champalimand is really the expert in our peer industry, unconsciously he built up this extremely exquisite space.

其实让我记起的倒不是这些，因为它们确实挺柔和地融在一起，不显眼，只是职业习惯才会关注到。次日到处走走，从山下到山上看了它各种房间，还有Spa（专业，可惜未及去享受）。绿树成荫，成林，让这些粉色西班牙建筑散落在其中，静中活泼，流连忘返。当然，在这里三天，白天尽情享受它的阳光、花园、椭圆泳池（一个人游泳），晚上也体验大厨Wolfgang Duck打理的全套餐，太丰富了，一菜式配一种酒，“酒不醉人，人自醉！”深深体味了（三天的）“小贵族”生活，撑死了，一顿晚餐吃了四小时！

Actually something I can remember is far from this as they indeed gently merged together. It is not easy to be noticed and only the man who are usually concerned about arts out of vocational habit will notice this. Walking around here in the next day, I could see all kinds of rooms from the foot of the mountain till rooms on the mountain, also spa (quite professional and what a pity that I failed to enjoy). Green tree became a great shade and formed a natural shade. These pink Spanish architecture scattered among of these, lively in the quiet land and you would even forget to return. Of course, during 3 days here, I could enjoy the sunlight to my heart's content, Garden and ellipse swimming pool (I can swim alone) and experience the good food offered by Chef Wolfgang Duck. The food here was quite rich and abundant. One dish collocated with a kind of wine. One saying goes like this, "wine does not intoxicate and people will be easy to get drunk!" We deeply experienced (3 days') "noble" life, so full, and it took me 4 hours to finish just one meal.

粉色，性感而青春，让人难忘！

Pink, sexy and youthful, you will never forget!

MANDARIN ORIENTAL
SAN FRANCISCO
旧金山文华东方酒店

14 ★★★★★

MANDARIN ORIENTAL, SAN FRANCISCO

Address : No.222 Sansome Street, San Francisco, CA 94104
Telephone : +(415) 276 9888
Http : //www.mandarinoriental.com.cn/sanfrancisco//

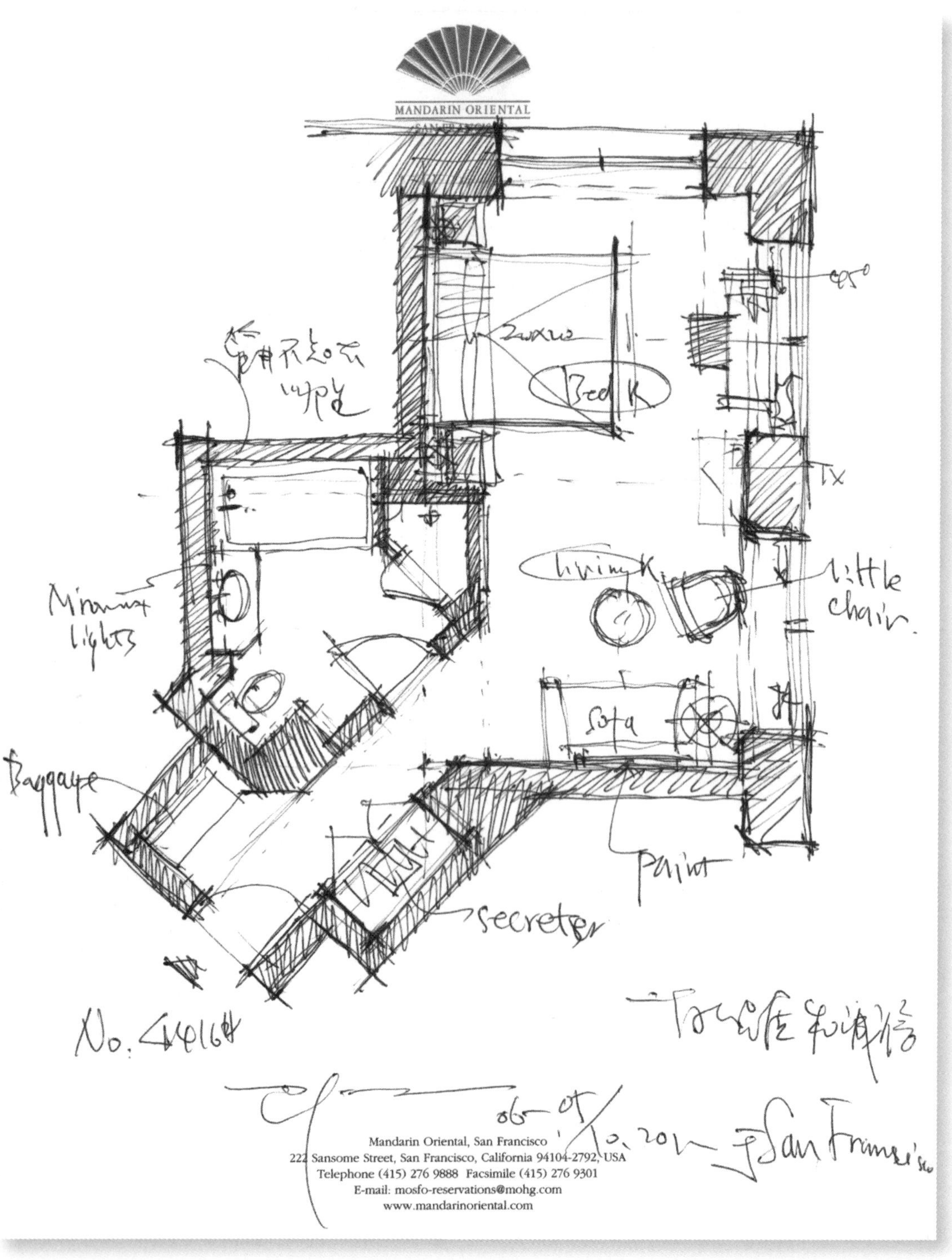

MANDARIN ORIENTAL
Bed R
Living R
little chair
Sofa
Paint
secreter
Baggage
TV
Mandarin Oriental, San Francisco
222 Sansome Street, San Francisco, California 94104-2792, USA
Telephone (415) 276 9888 Facsimile (415) 276 9301
E-mail: mosfo-reservations@mohg.com
www.mandarinoriental.com

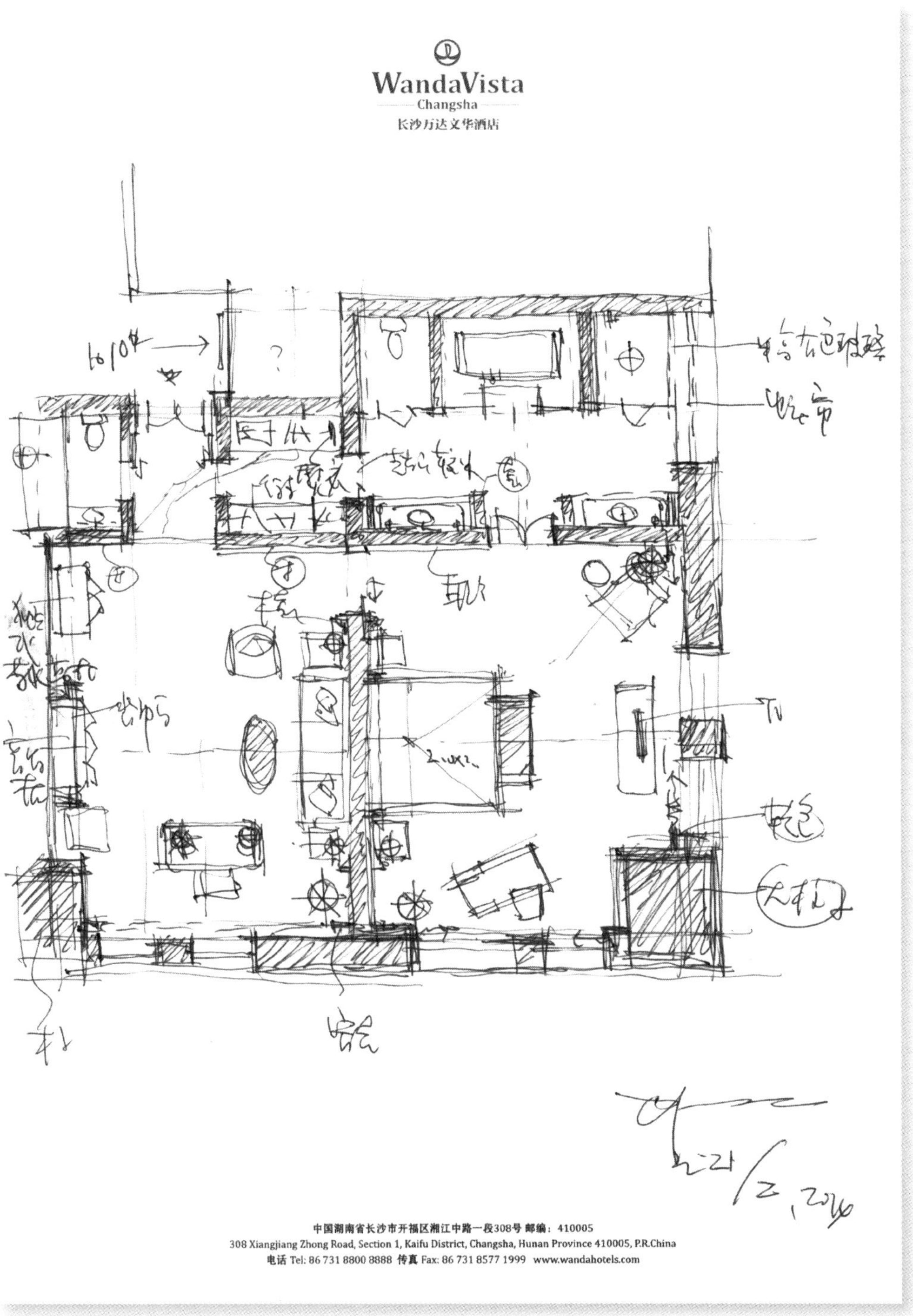
WandaVista
Changsha
长沙万达文华酒店
TV
中国湖南省长沙市开福区湘江中路一段308号 邮编：410005
308 Xiangjiang Zhong Road, Section 1, Kaifu District, Changsha, Hunan Province 410005, P.R.China
电话 Tel: 86 731 8800 8888 传真 Fax: 86 731 8577 1999 www.wandahotels.com

WandaVista
Changsha
长沙万达文华酒店

长沙万达文华酒店

15 ★★★★★

WANDAVISTA,CHANGSHA

Address : No.308 Xiangjiang Road, Section 1,Kaifu District, Changsha,Hunan Province 410005,P.R.China
中国湖南省长沙市
开福区湘江中路一段308号

Telephone : +86 731 8800 8888

Fax : +86 731 8577 1999

Http : //www.wandahotels.com/

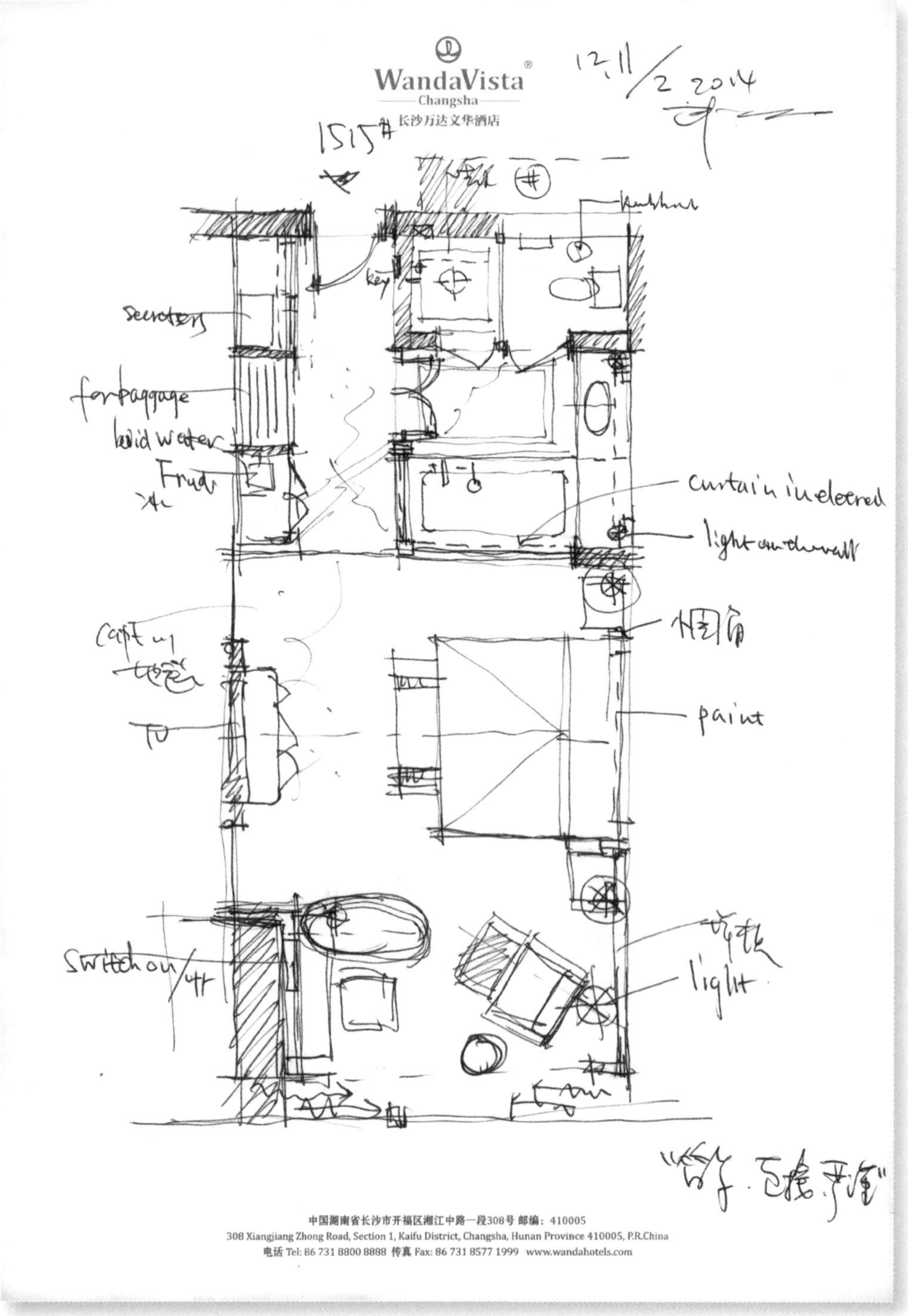
WandaVista
Changsha
长沙万达文华酒店
12/11/2 2014
1515#
key
for baggage
boil water
Fruit
curtain in electred
light on the wall
paint
TV
switch on/off
light
中国湖南省长沙市开福区湘江中路一段308号 邮编：410005
308 Xiangjiang Zhong Road, Section 1, Kaifu District, Changsha, Hunan Province 410005, P.R.China
电话 Tel: 86 731 8800 8888 传真 Fax: 86 731 8577 1999 www.wandahotels.com

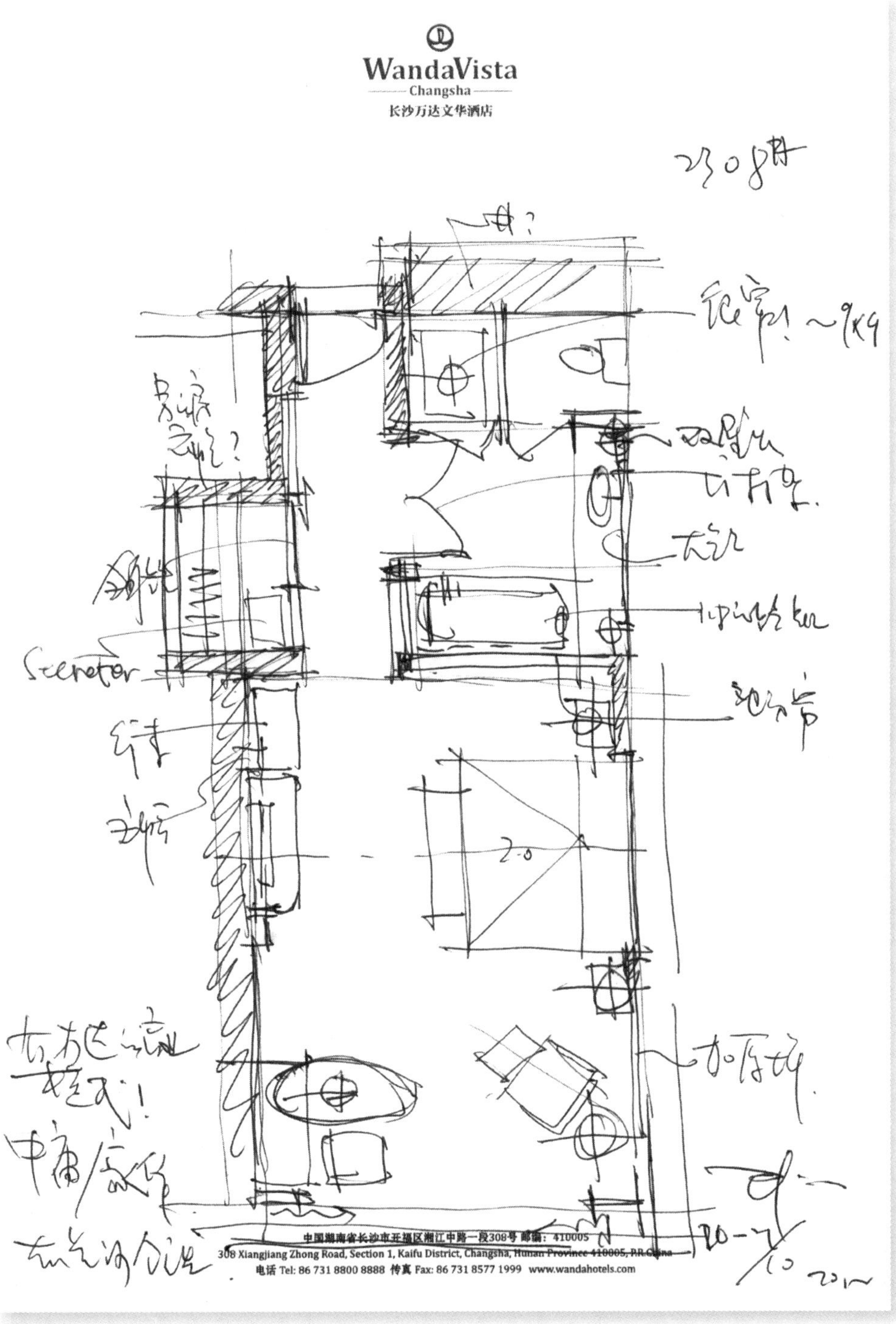
WandaVista
Changsha
长沙万达文华酒店
Secretor
中国湖南省长沙市开福区湘江中路一段308号 邮编：410005
308 Xiangjiang Zhong Road, Section 1, Kaifu District, Changsha, Hunan Province 410005, P.R.China
电话 Tel: 86 731 8800 8888 传真 Fax: 86 731 8577 1999 www.wandahotels.com

21–22/12. 2014

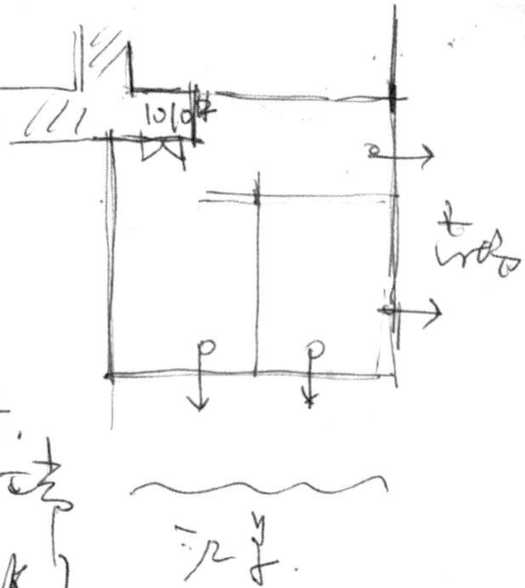
1010#

占了便宜了!

占了长沙的大便宜了!

因为订在网上 878元/天的大客房,

到达已经是十点钟左右了,晚上,请check in,前台说已经免费升级了。

我们进去看可能也是像这样的那里没有,哈!

江景豪华一整层的房间,1010#,十楼!

去到走道尽端已经觉得不是一般的入口,尽端的入住者。

一推门,我们又是大客厅,一厅之中又有一台,可能是……还不识之中,大了,大厅近二十平方米的……能看江景……

但又没有把它们当作一般的大便宜了,我们也是住了的傻的,不知道下回……太好,要回房去竟是了。希望这种好事能持续下去。

酒店也很好,大堂……中一

中国湖南省长沙市开福区湘江中路一段308号 邮编:410005
308 Xiangjiang Zhong Road, Section 1, Kaifu District, Changsha, Hunan Province 410005, P.R.China
电话 Tel: 86 731 8888 8888 传真 Fax: 86 731 8572 1999 www.wandahotels.com

长沙万达文华酒店 *WandaVista, ChangSha*

Gain Extra Advantage
占便宜了

吃亏却占了大便宜了！

I gained an extra advantage for losing!

因为有应酬，878元/天的大床房，到店已经是十点多十一点了。晚上，正常 cheek in，前台说已经免费升级了。我心里想可能也是像之前的习惯说法，给一个江景或者景观好一点的房间，1010号，十层！去到走道尽端已经觉得不妥——双门入口，尽端凹入位置。

Because of a dinner party, it's nearly 11 o'clock when I arrived at the hotel. I booked a king size bed room which was cost 878 yuan per day. The reception told me they had already upgraded my booking when I checked in. I thought they might normally arrange a room with river view or other good scenery outside. My room number was 1010. It was on 10th floor! Walking to the end of passage, I felt not good: The room entrance was double door, and it was deep inside the wall.

一推门，我的天，是大套房，一层之中最尊贵的一间。可能之前专门和同事来“体验”的豪华还不及之一半。大厅、大房，近二十多平方米的主洗手间，转角江景……

I pushed the door and found it was a suite! Oh my god! It was the best room on that floor. The suite that I especially came for "experience" with my colleague last time was much less deluxe than this one. There was a big living room, big bedroom and a master bathroom in over 20 square meters, and I could see the river for 270 degrees…

很久没有捡到这么爽的大便宜了。或者也是住多了的缘故，又或者万达生意太好，普通房都卖光了。希望这种好生意继续下去。

I haven't stayed in a good suite with such a low price for long time. Maybe that's because I used to stay here for many times and became VIP for them, or because all common rooms were full booked. due to the unbrisk trade of Wanda I hope Wanda could keep such good business.

酒店是我家，兴隆靠大家！

Hotel is like a home, and we support its business together!

11/2. 2016

"因为"

因为信任，所以才有更多的激情。今天，我们到第四家的万达"第一"的宾馆，感到万达酒店的发展迅猛前进。又一次是住进长沙开福区。虽然因为技术原因我们不能接受很多之合作，但不妨碍对万达的佩服之情。

再次见证我们的酒店签约，一年内又增加了许多许多，分别遍布全世界的城市，所占的领域也是成功之前提。但是很少做到。

所以这次的"签签字"又多了一些东西，而且更加关注万达酒店集团的发展！

入住时前台的服务员的亲切交流和询问，入住后感受细节，早餐很好……，感受到万达人对细节的重视，好棒的人员服务。现代的企业们最缺——人才。万达有着很多最大的财富。相信他会成为世界最好的酒店集团!!

中国湖南省长沙市开福区湘江中路一段308号 邮编：410005
308 Xiangjiang Zhong Road, Section 1, Kaifu District, Changsha, Hunan Province 410005, P.R.China
电话 Tel: 86 731 8808 8888 传真 Fax: 86 731 8857 1999 www.wandahotels.com

Because
“因为”

因为住多了，所以少了画的激情。入房后，翻看了第四季的万达“万一”的季刊，看到了万达酒店系统的发展迅猛势头，又一次关注这家地产集团。虽然因为投标方式，我们不接受而放弃与之合作，但不影响对万达的佩服之情。

Because I used to stay this hotel for many times, there is not many things exciting for me here. After I had entered hotel room and red the fourth quarterly named WAN YI from Wanda Group, it caused my attention to this property group again because of the fast development of Wanda's hotel system. Although we had given up the cooperation with them because we could't accept to bid, that won't affect my appreciation to Wanda.

再看看他们的酒店名字，一年内又增加了许多许多。市场是最好的证明。所占的资源就是成功的前提。微信、微博、阿里系列，占的就是人气。

Look at their hotel list, which became much longer within a year. Market is the beat proof. How many resource you take decides your success. Such as Wei Chat, Micro blog and Alibaba, they all take popularity.

所以这次就是“老老实实”又画了一次平面，而且更加关注其成功“简单”的招数！

So this time I drew the layout "honestly" again, and pay paid more attention to the simple but mature details!

入住时，前台的靓女亲切地交谈和询问入住了几次感觉如何，早餐好不好……感受到其人才的成熟，架构的人亦成熟。现代的企业什么最值钱——人啊，人才。这才是万达储备最大的财富。相信他会成为世界级的酒店集团！

During I stayed here, the lady in reception talked to me friendly, asking me about my feeling about staying for times and about breakfast... That made I feel the mature structure of staff of Wanda. What is the most valuable for modern enterprise? Persons of ability, of course. It's the most valuable wealth of Wanda. I believe they will become a international hotel group!

ST REGIS
SANYA · YALONG BAY
三亚亚龙湾瑞吉度假酒店

三亚亚龙湾瑞吉度假酒店

16 ★★★★★

STREGIS, SANYA YALONG BAY

Address : Yalong Bay National Resort District,Sanya, Hainan,China 572016
中国海南三亚亚龙湾国家旅游度假区

Telephone : +86-898-8855-5555

Fax : +86-898-8855-5555

Http : //www.sanya-regis.com/

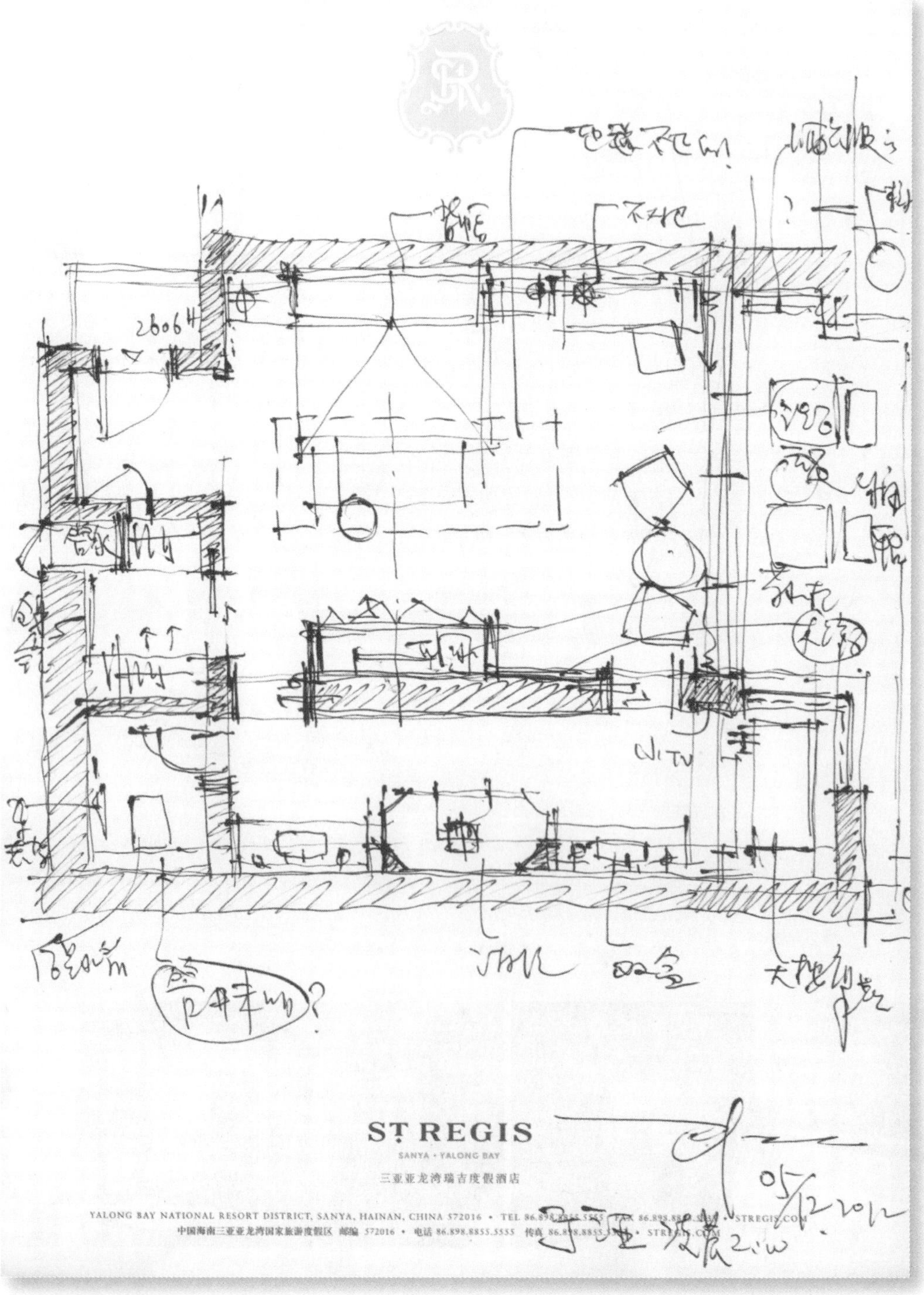

2606
ST REGIS
SANYA · YALONG BAY
三亚亚龙湾瑞吉度假酒店
05/12 2012

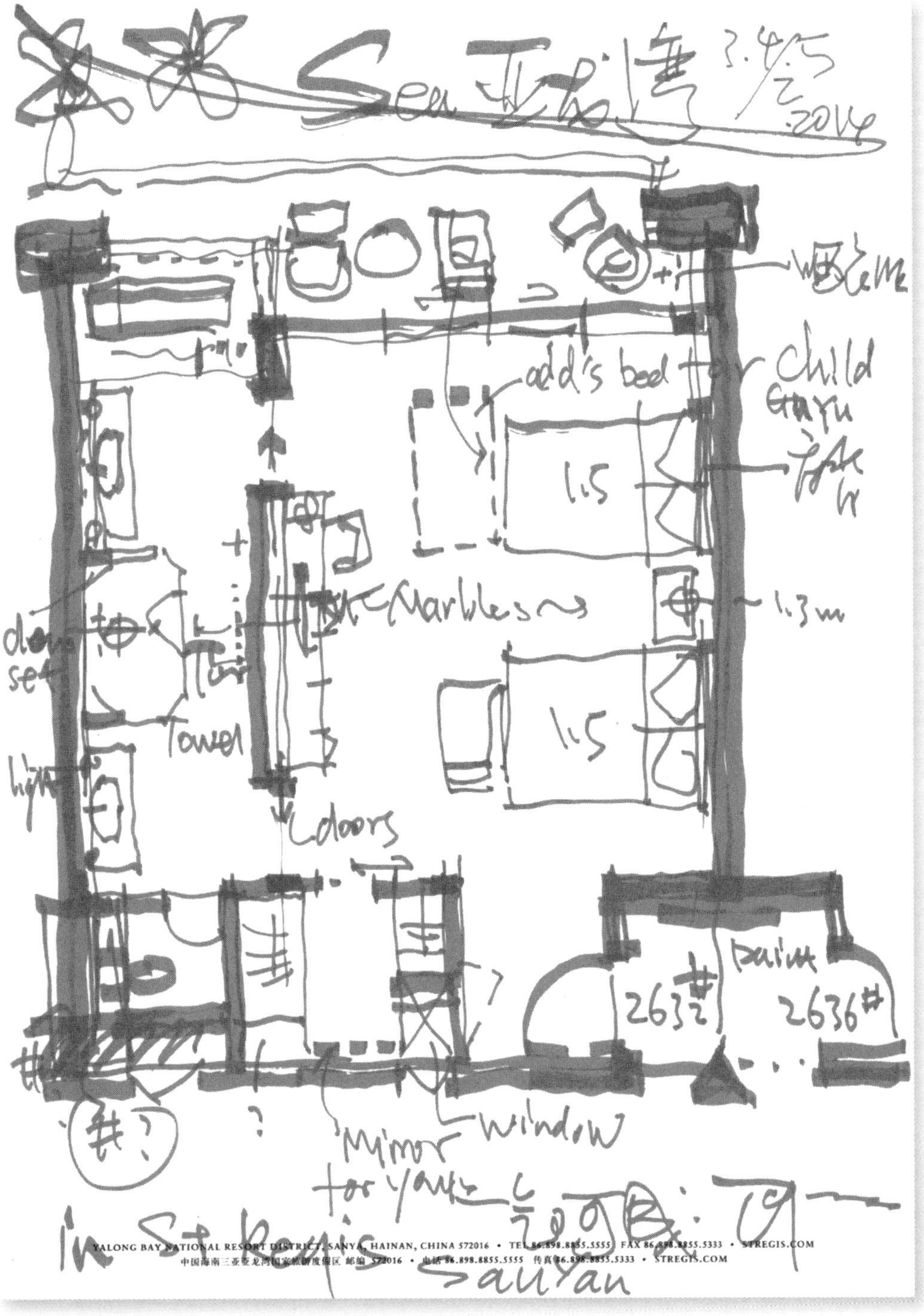
Sea
3.4/5
2014
add's bed for child
1.5
Marbles
1.3m
Tower
doors
1.5
2632#
2636#
window
Mirror
for yan
in St. Regis
Sanyan
YALONG BAY NATIONAL RESORT DISTRICT, SANYA, HAINAN, CHINA 572016 • TEL 86.898.8855.5555 FAX 86.898.8855.5333 • STREGIS.COM
中国海南三亚亚龙湾国家旅游度假区 邮编 572016 • 电话 86.898.8855.5555 传真 86.898.8855.5333 • STREGIS.COM

晒阳台

去三亚享阳光，吃海风，但如果是公干出差，那就没有这等休闲福气了。第一次住亚龙湾瑞吉酒店，到达时已近半夜，约了朋友在大堂吧里聊聊天，饿得已太慌，只好拿房间的小冰箱的小吃、饮料胡乱吃了一下，浴缸，没用过；阳台没到过；它的餐厅没试过，更没有用到健身套房（应当说是服务的客厅），浴缸也只用了一半，又拿了这些东西，倒是因为太匆忙，没有拿短袖衣服，第二天早餐时真想把长衫脱掉，太热了！

第二次又住这些年假期，无所事事，早上睡到九、十点，阳光太猛，阳台家具也晒得滚烫，根本坐不下，倒是近日落时可以休闲地在阳台喝茶，看看来去，望望西边，看着海，吃着风，再随意地摆弄着阳台的家具，组合出最懒的方式。第三次变聪明了，早点起来，趁着晨子，可以晒着还没有烫人的阳光，听着音乐收音机，懒洋洋地斜躺着，半眯着，那才是真正的晒阳台，在阳台上晒，舒服！倒是房间的样子慢慢模糊了……

早知道加床这么贵（小孩子的），那我可以在阳台上过了，可以好好感受夜里晒月光，晒星光的闲情，天晴，入夜可以数一下星星。

三亚亚龙湾瑞吉度假酒店 *Stregis, Sanya Yalong Bay*

In the Balcony
晒阳台

去三亚享阳光，吹海风，但如果是公干出差，那就难有这等休闲福气了。第一次住亚龙湾瑞吉酒店，到达时已近半夜，约了朋友在大堂吧想聊聊天，结果已打烊，只好拿房间小冰箱的小吃、饮料胡乱吃喝一下。浴缸没用过；阳台没到过；衣帽间没动过，更没有用到过传菜窗（应当说是服务的窗口），洗脸盆也只用了一半，双盒之中的一个。倒是因为太匆忙，没有拿短袖衣服，第二天见客时真想把长袖脱掉，太热了！

Go to Sanya to enjoy the sunshine and sea breeze. If you are on business trip, it is difficult to have this leisure mood. So it is never easy to have such kind of feeling. It is the first time for me to live in Stregis on Sanya Yalong Bay. When I arrived, it was already midnight. I had a date with my friend and had a chat in the lobby bar. However, the lobby bar was closed and we had to go back room to have some snack or take something to drink from the refrigerator. I had never used the bathtub. I had never been to the balcony as well. I have never touched the clothing room and never been to the service window. I just use half of the washing basin, one of the two boxes. I did not take the short-sleeve clothes as I was too busy. I really planed to remove my long-sleeve when meeting the customer tomorrow as it was really too hot.

第二次入住是过年度假，无所事事，早上睡到九十点。阳光太猛，阳台家具热得烫屁股，根本坐不下，倒是近日落时可以休闲地在阳台喝茶，看看杂志，写写画画，看着海，吃着风，肆意地摆弄着阳台的家具，组合出最懒的方式。第二次变聪明了，早点起来，光着膀子，可以晒着还没烫人的阳光，听着手机收音机，懒洋洋地斜躺着，半眯着。那才是真正的晒阳台，在阳台上晒，舒服！倒是房间的样子慢慢模糊了……

The second time for me to live here is when I having holiday during the Spring Festival. I had nothing to do and slept till 9 to 10 o'clock in the morning. Too strong sunlight and my furniture in the balcony got very hot and I could not sit down at all. We had to choose the sunset time to drink some tea leisurely at the balcony, see the magazine and draw some picture, overlook the sea, blow the sea breeze, set up the furniture in the balcony, combine all these together in the laziest way. The second time I got clever. I got up early in the morning with my shoulder nude. I could sit in the mild sunlight, listen to the radio program from the mobile. Lean lying there lazily, closing the eyes, that is my space in the balcony and I felt so comfortable. I even forgot what the room look like…

早知道加床这么贵（小孩子的），那我就在阳台上过了，可以好好感受夜里晒月光、晒星光的闲情。天晴，入夜可以数一下星星。

If I knew before that adding the bed would be so expensive (for the children), I would rather sleep on the balcony and enjoy the moonlight at night and find the leisure moment under the starry night. Early in the morning, I can look up the starry night and count the hundreds of thousands of stars.

懒，想出来的！

春节长假，无所事事，看看天，看看海，看看微信。近日，IT大亨莫过于这个公司搞出来的"红包"了。合法不？可以随意支付，有监督不？忙坏了把人。

看看我们的酒店、商店世界，人气不过如此，并不像这种针对"懒惰"而生的公司风生水起。

电商——人们不出门就可以生活，没什么感受的无品质生活。

微信——人们动动手指就可知天下事，不知真假的世界。

网银——人们不去银行就可串行交易，银行的精英都去做后台了。

那消失之人还要什么，好像什么也不需要了。人都可以在家玩了。试着放下微信，不看不行，越看越依赖。小马哥的心理研究确实厉害，渐渐控制你我的生活，一家伟大的公司要不要讲道德呢？

懒懒的设计也在营造懒懒的状态，致富可不是，我将设计也分成几百个部落，每部只收三两块，这样就可以消除动辄上千的自行设计费，大家就不敏感了，像网上的支付一样，小小的金额就会让你解放了！

将设计变成方便的必需品，就像每天要吃、出门要住一样，那我们做设计就可以更专心、专一，也能过好的生活，台时候在热：阴招就是好招，晴招也赞！

一圈人聚聚，手机不离手，爱摇的人爱微信，爱设计的我还是坚持放下手机，晒晒太阳，更写意一些，特别是在这么好的酒店。

Lazy and Think about

懒，想出来的！

春节度假，无所事事，看看天，看看海，看看微信。近日，IT大事莫过于这个公司搞出来的"红包"了，合法不？可以随意支付。有监督不？忧天之杞人。

In Spring Festival, I had nothing to do but looked at the sky, watched the sea, and read the Wei Chat. Recently, the biggest issue in IT area should be the " red pocket" from Tencent. Is it legal? The red pocket can be paid easily. Is it under control? Somebody worried too much.

看看我们住的酒店，虽说过节，人气不过如此，并不像这种针对"懒惰"而生的公司风生水起。

Look back to the hotel we stayed. It's in festival vacation, but there are still a few guests. Not as popular as Tencent, which is a company that services for "lazy persons".

电商——人们不出门就可以生活，没什么感受的无品质生活。

Electronic business makes people live a low quality life without going outdoor.

微信——人们动动手指就可以知天下事，不知真假的世界。

Wei Chat makes people know everything without distinguishing true or false by just moving the fingers.

网付——人们不去银行就可以串行交易，银行的精英都去做后台了。

Online payment makes people finish trade without going to bank, so the bank elite become backstage supporters.

那活生生的人还要什么？好像什么也不需要了。人都可以虚拟了，试着抗拒微信，不看不行，越看越依赖。小马哥的心理研究确实厉害，渐渐地控制我们的生活。一家伟大的公司要不要讲道德呢？

So what else do people need? They seem to need nothing else, and become virtual character online. How can you refuse Wei Chat? You can't. But the more you read, the more you dependent on it. The psychology research of Mr. Ma is really perfect. He controls our life day by day. As a great corporation, is it necessary to have strict morality?

懒懒的设计也应当享受懒懒的状态，现实可不是。或将设计分成几百个步骤，每步只收三两块，这样就可以避免设计费动辄高达每平方米上千元。大家就不敏感了，像网上的支付一样，小小的金额就会让你麻木了！

The one who designs for lazy life should also enjoy lazy life himself. But reality is not like that. Or we can divide our design into hundreds of steps, and each one charges only two to three yuan. In that case, the design fee won't be nearly 1000 yuan per square-meter, and our client won't be so sensitive with price. Just like the online payment, paying small amount of money each time makes you feel nothing!

将设计变成方便的必需品，就像每天要吃，出门要住一样，那我们做设计就可以更专心、专一，也能过上好的生活。有时候在想：阴招就是好招，明招必败！

Turn design into convenient necessity like food for everyday or hotel for trip, so that our job can be just focus on design, and we will have better life. Sometimes I think: Plot is good for business. The normal method will be failed.

一圈人聚聚，手机不离手，爱炫自己的爱微信。爱设计的我还是应该放下手机，晒晒太阳，更写意一些，特别是在这么好的酒店。

In a party, people always hold their mobile phones and like showing off their Wei Chat post. As a person loving design, I should release my mobile, enjoy the sunshine, and be leisure, especially when I am in such a good hotel.

THE LANGHAM
Shenzhen

深圳东海朗廷酒店

17 ★★★★★

THE LANGHAM, SHENZHEN

Address : No.7888 Shennan Boulevard
Futian District, Shenzhen
518040,China
中国深圳市福田区
深南大道7888号
Telephone : +(86) 755 8828 9888
FAX : +(86) 755 8828 9888
Http : //shenzhen.langhamhotels.com/

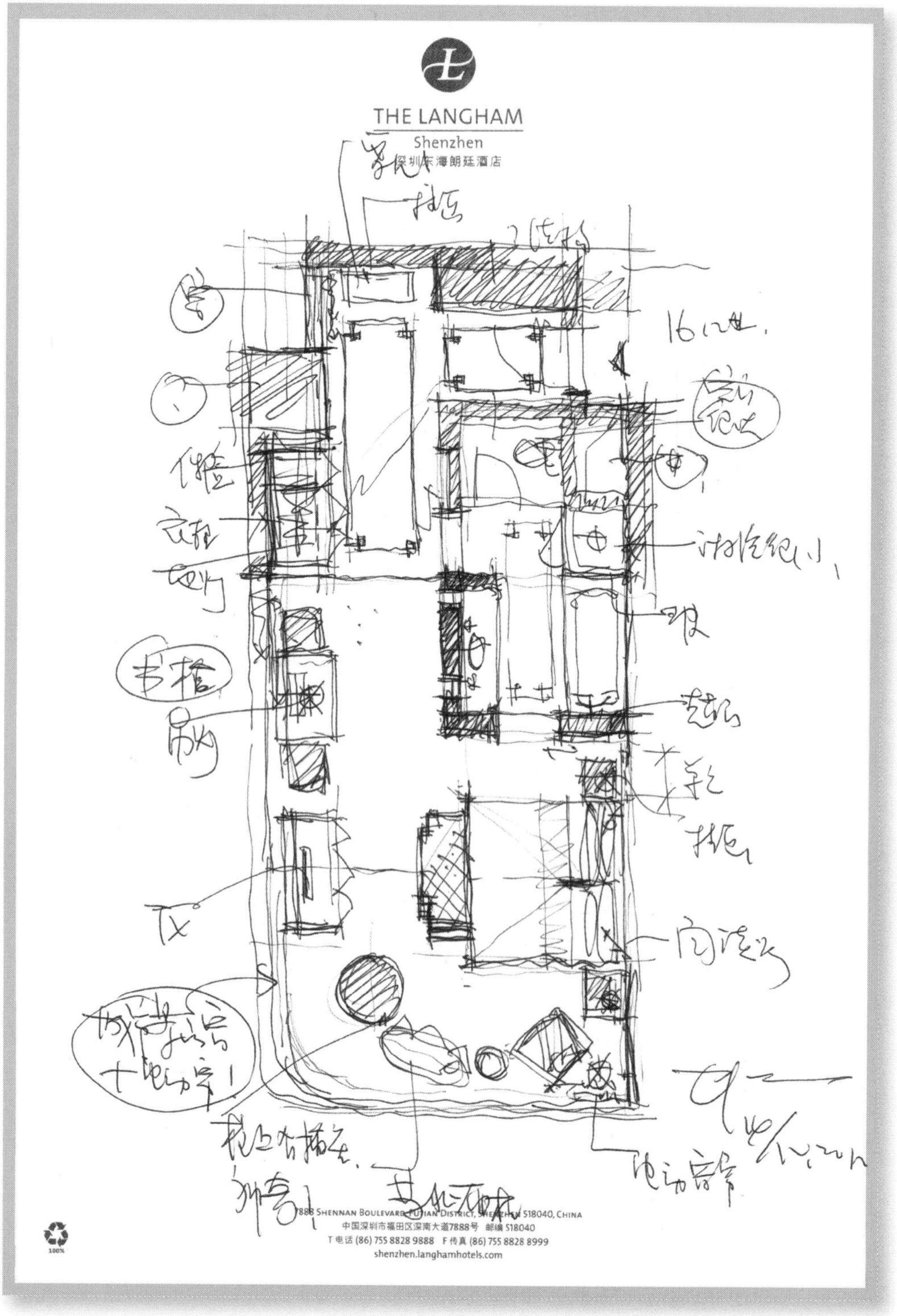

THE LANGHAM
Shenzhen
深圳东海朗廷酒店
7888 SHENNAN BOULEVARD, FUTIAN DISTRICT, SHENZHEN 518040, CHINA
中国深圳市福田区深南大道7888号 邮编 518040
T 电话 (86) 755 8828 9888 F 传真 (86) 755 8828 8999
shenzhen.langhamhotels.com

Yes!（耶，西！）

听说深圳开了一家朗廷酒店，之前也住过上海新天地的一品朗廷酒店，感觉不错。但提起它，心里想，朗廷不是一个传统英国品牌吗。它的酒店一般都是古典和传统、老牌的吗，像伦敦的那间！

带着疑问去了深圳，哗，还真在一幢新建筑物里，会失去方向吗？

还好，是一间高端的房间，安静、通透，大转角房间，双边采光，关键是新派、清雅；同时的"新"英式设计，细节而又惊喜，难得在国内遇到如此优雅的"英伦淑女"，果然是大品牌。印象很深的几个部位：方形的书/餐桌，一盏台灯，家的意境；独立式椭圆形浴缸，完美无瑕，收放自如；全触控一体化的电动帘，灯光、隐私都能做到控制，达到各方面的适合，……

难得的精美，给深圳的酒店添上精彩细腻的一笔。清新，我说：Yes！不错的英伦风！

Yes! Jesse

Yes!耶西

听说深圳开了一家朗廷酒店，之前也住过上海新天地的同一品牌酒店，感觉不错。但现代的，心里想，朗廷不是一个传统英国品牌吗？它的酒店一般都是古典和传统、奢华的吗，像伦敦的那间！

I heard that one Langham Hotel was opened in Shenzhen, before I had ever lived in the same brand of this hotel in Shanghai New Paradise and felt good. It is quite modern. Isn't Langham one traditional British brand? Is this hotel classic and traditional or luxurious? Or it is like that one in London?

带着疑问去了深圳。哗，还是在一幢新建筑物里。会失望而归吗？

With such a question, I went to Shenzhen. Well, it is still in one new building. Shall I return with disappointment?

还好，是一间尽端的房间，安静、迂回，小转角房间，双边采光。关键是新派、清雅；白调的"新"英式设计，注重细节而又给人惊喜。难得在国内遇到如此优雅的"英伦淑女"，果然是大品牌。印象很深的几个部位：方形的书/餐柜、一盏吊灯、家的意境；独立式椭圆形茶水柜完美无瑕，收合自如；全弧位一体化的电动帘，观景、隐私都能自如控制，还有走廊的条窗与挂画的组合……

It is just ok. It is one room at the end, quiet, circuitous, room at small corner, and both sides are good lighting. The most important thing is refreshing and elegant, white background "new" British way of design, details with surprise. It is difficult to meet such an elegant "British girl". It is indeed big brand. Several parts that left me deep impression: the square book/ dinning cabinet, a lamp, family atmosphere, independent oval tea cabinet: perfect and flexible, all-arch integrated B/O curtain, the view and private can be controlled flexibly by your own with the perfect combination between the aisle window and the hanging picture.

难得的精美，给深圳的酒店添上特殊调子的一员，清新。我说：Yes!不错的英伦风！

It is the rare exquisite art, which adds the special tone to the Shenzhen Hotels. It is just refreshing, so I said :Yes! Very nice British fashion!

深圳东海朗廷酒店 *The Langbarn, Shenzhen*

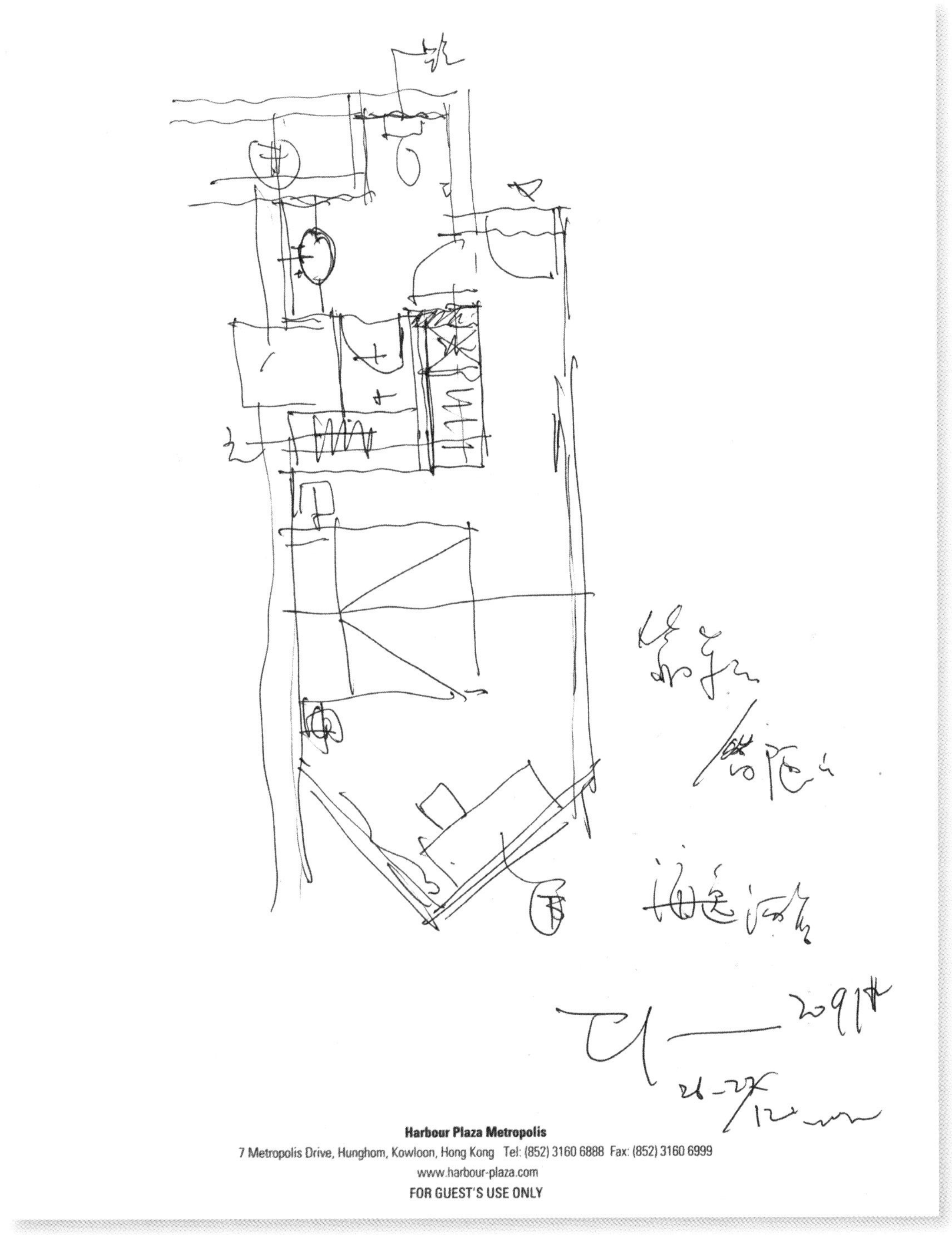
Harbour Plaza Metropolis
7 Metropolis Drive, Hunghom, Kowloon, Hong Kong Tel: (852) 3160 6888 Fax: (852) 3160 6999
www.harbour-plaza.com
FOR GUEST'S USE ONLY

HARBOUR PLAZA
METROPOLIS
KOWLOON · HONG KONG
香港都会海逸酒店

18 ★★★★

HARBOUR PLAZA METROPOLIS, HONGKONG

Address : No.7 Metropolis Drive
Hunghom, Kowloon
Hong Kong
香港九龙红磡都会道 7 号

Telephone : +(852) 3160 6888
FAX : +(852) 3160 6999
E-mail : enquiry.hpme@harbour-plaza.com
Http : //www.harbour-plaza.com/

Shangri-La hotel
BAOTOU
包头香格里拉酒店

19

★★★★★

SHANGRI-LA HOTEL, BAOTOU

Address : No.66 Min Zu East Road
Qing Shan District,
Baotou, Inner Mongolia,
014030, China
中国内蒙古自治区包头市青山区民族东路66号
Telephone : +(86 472) 599 8888
FAX : +(86 472) 599 8999
Http : //www.shangri-la.com/

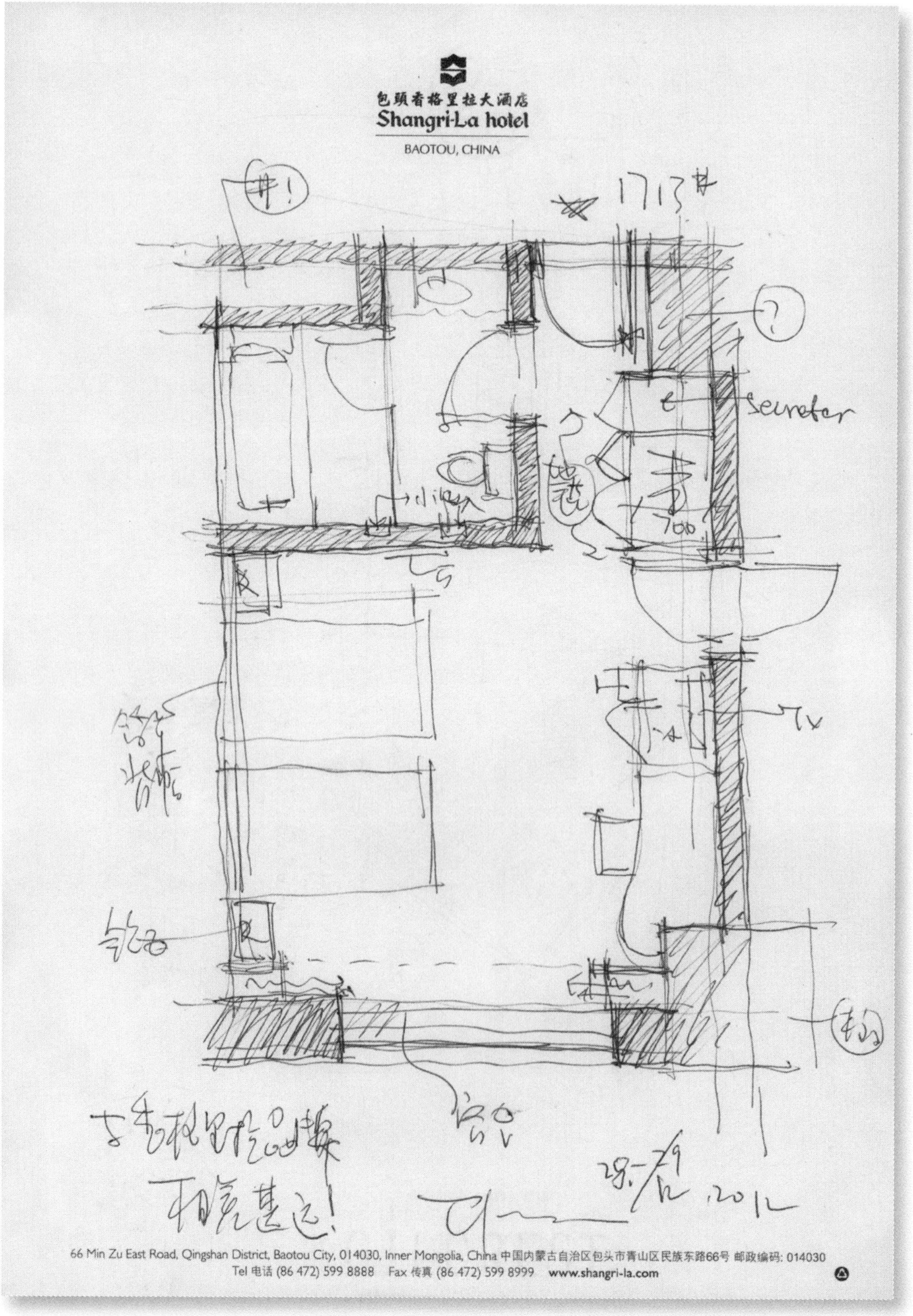

包頭香格里拉大酒店
Shangri-La hotel
BAOTOU, CHINA
Secretar
TV
66 Min Zu East Road, Qingshan District, Baotou City, 014030, Inner Mongolia, China 中国内蒙古自治区包头市青山区民族东路66号 邮政编码: 014030
Tel 电话 (86 472) 599 8888 Fax 传真 (86 472) 599 8999 www.shangri-la.com

小城市，大作为。

没想到在"小城市"包头有这么经典的品牌，也证明大品牌的进入对城市的酒店品质提升的效应。

因为包头的酒店近日开去参观当地的近十间酒店，也有五星级的，但绕了一圈下来还是很自然地选择香格里拉酒店下榻。

一者品牌，不担心小城市找较差的；

二者地段不错，方便，因是很冷的初冬了；

三者正好体验在小城市里的大品牌有什么不同的策略和定位。

房间很宽，很短，先天的不足，或者当初只考虑大床房吧，只是我们两个"老男人"入住标准间，单房面积现在看来肯定是不达标的。投入也一般般，当然高级就没什么好说的，只能说算是"入乡随俗"，不用太争就能领先同行，但生意相当好！让我们的客户对在包头开店经营有了希望和信心！

当然，一贯精美的早餐也为它最后加了分！

品牌的力量，在小城市也可以！

Small city, big ambition
小城市，大作为

没想到在“小城市”包头有这么经典的品牌，也证明大品牌的进入对城市的酒店品质有提升的效应。

I never expect in the “small city” like Baotou still have such a classic brand, which also proved that the introduction of the big brand can greatly enhance the brand reaction.

因为包头的酒店项目而去考察当地的近十间酒店，也有五星级的，但匆匆一圈下来还是很自然地选择香格里拉酒店下榻。

I went to Baotou to conduct survey on nearly 10 hotels there due to the hotel project in Baotou. Some are Five Star, however, hurry for a round and find that I still love to stay in Shangri-La Hotel.

包头香格里拉酒店 *Shangri-La Hotel,Baotou*

一者，品牌，在不熟悉的城市找稳妥的；

Firstly, brand, find the safe and secure brand due to unfamiliar with this city.

二者，地段不错，方便。因是很冷的初冬了；

Secondly, nice section and convenient as it is quite cold in early winter.

三者，正好体验在小城市里的大品牌有什么不同的策略和定位。

Thirdly, it is also one good opportunity to experience the difference of big brands in small city in terms of strategy and positioning.

房间很宽、很短，先天的不足，或者当初只考虑大床房吧。只是我们两个“老男人”入住才显拙。单房面积现在看来肯定是不够的。投入也一般般，当然效果就没什么好说的，只能说真是“入乡随俗”，不用太争气就能领先同行。但生意还是相当的好！让我们客户对在包头开店经营有了参考和信心!

The room is quite spacious but short, congenital deficiency, or maybe at that time just consider the big bed. We, just "two old men", find it quite narrow when staying. Space in the single room is definitely not enough. The facility is just so so. Of course, we can not have further comment on the effect. We can just say "following the custom". Do not have to be too critical and competing with your counterpart and you will find that you still take leading position in your peer industry. However, hotel business in Baotou is still quite good! This makes our customers in Baotou who are planning to run a hotel like that have more reference and confidence!

当然，一贯精美的早餐也为它最后加了分!

Of course, one meal of exquisite breakfast adds it the high scores!

品牌的力量，在小城市也可为!

Power of the brand, small city can make a good business as well!

FOUR SEASONS
Hotels and Resorts
香港四季酒店

20 ★★★★★

FOUR SEASONS HOTEL, HONGKONG

Address : No.8 finance street, central,Hong Kong
香港 中环金融街8号

Telephone : +(852) 3196 8888

FAX : +(852) 3196 8899

Http : //www.fourseasons.com/

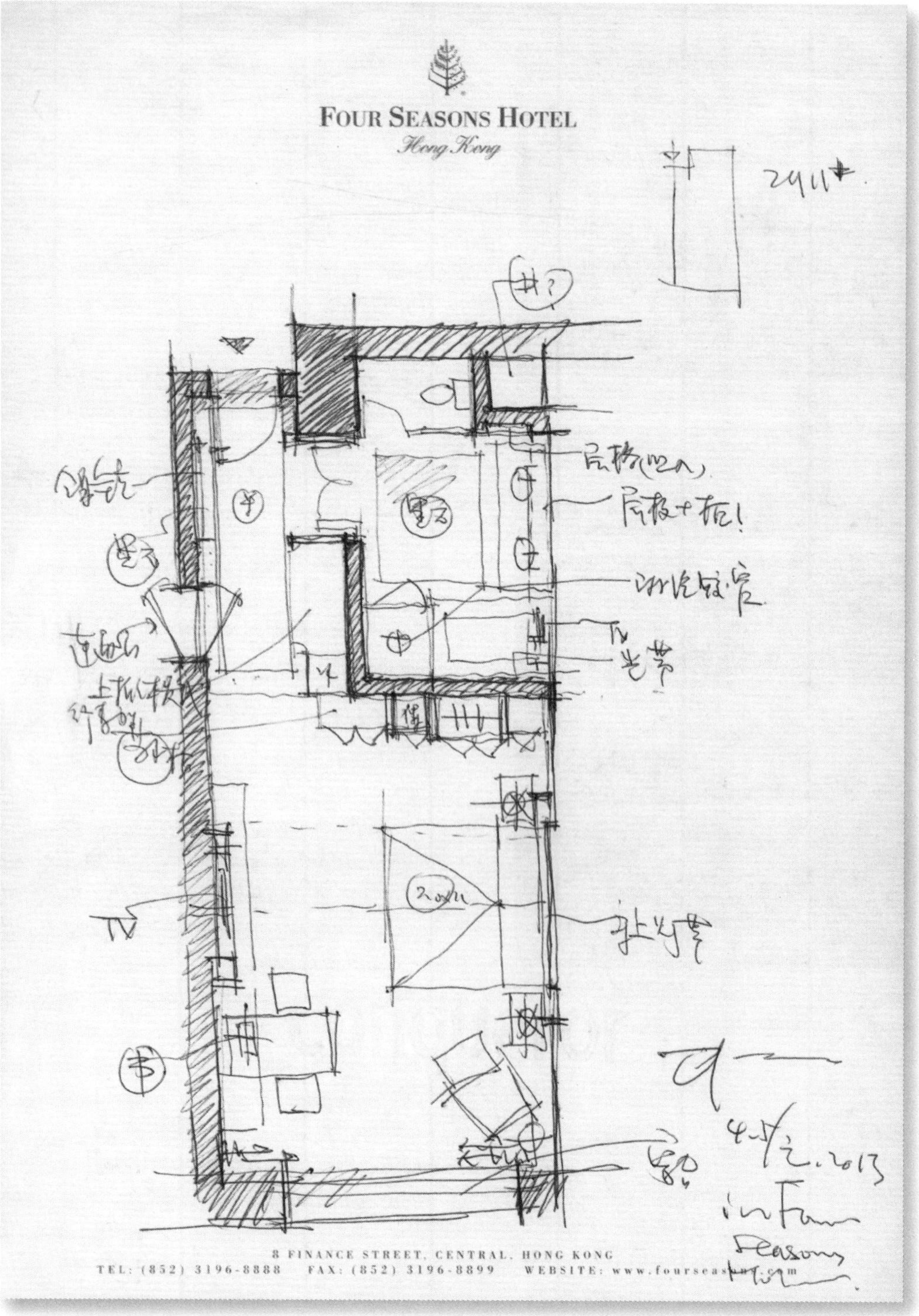

FOUR SEASONS HOTEL
Hong Kong
8 FINANCE STREET, CENTRAL, HONG KONG
TEL: (852) 3196-8888 FAX: (852) 3196-8899

品牌，是你最好的选择

闭着眼睛再次选择了它，香港的四季酒店，那里有家的感觉，温馨感。倒是对所谓的艺术型、设计型酒店不感兴趣，或是我的偏见，下次补一下，姑且住一下，希望有不同的感受！

品牌是什么？不是一天两天就能建立的，有积累，有历史，有故事，当然关键是有认同才行。在不知如何选择的情况下，保险或是一种策略，而品牌就是最安全的选择。其实反过来说，四季的布局不见得完美，有几处还是搞不懂为什么要这样而不是那样处理，倒是一些细节，侍者，欢迎的字条，晚上的书，早上的报纸，惬意得让你忽略了内部的设计，这或者就是大品牌高明之处，将各种细节不经意地融合在一起，相信，你的选择就是最合适的！

香港四季酒店 *Four Season Hotel, Hongkong*

Brand is Your most Safe Choice
品牌，是你最安全的选择

闭起眼就再次选择了它。香港的四季酒店，配套有商场、海景房。倒是对所谓的艺术型，设计型酒店不感兴趣。或是我的偏见，下次补一下，权且住一下，希望有不同的感受！

Just close the eyes and you will choose it once again. Four Season Hotel in Hongkong , which is equipped with shopping mall, harbor view room. Personally I don't like the artistic and design-oriented type of hotel. Maybe this is my prejudice. I will add it in my next article, just to live and feel, hope that I can have different feel!

品牌是什么？不是一天两天就能垒起来的。有积累、有失败、有探索，当然关键是有认同才行。在不知如何选择的情况下，保险或是一种策略，而品牌就是最安全的选择。其实使用起来，四季的布局不见得完美，有几处还是搞不懂为什么要这样而不是那样处理。倒是一景遮百短，倚窗，白天看看景，晚上看看书，早上看看雾，惬意得让你忽略了内部的设计。这或者就是大品牌的高明之处，将各种优势不经意地糅合在一起。相信你的选择就是最合适的！

What is the brand? Brand can not be built up by just one or two days. Accumulation, failure, exploration, of course all brands should have a common point, that is, being highly recognized by the market. If you do not know how to choose from so many hotels, the most secure way or tactic is to choose the brand and this is the most secure option. Actually when we lived in Four Season, the layout is far from perfect. There are several points that I still do not know why they will process like that. Fortunately the good scenery helps cover all the defects: leaning on the window and seeing the view in the daytime, reading book at night, seeing the mist in the morning, you will quite enjoyable and pleasant and completely neglect the inner decoration and design. That may be the smart point of the big brand. Put each kind of advantage together unconsciously, I believe, your choice will be your best option!

MANDARIN ORIENTAL
SANYA
三亚文华东方酒店

21 ★★★★★

MANDARIN ORIENTAL, SANYA

Address : No.12 Yuhai Road, Sanya City 572000, Hainan, PRC China
中国海南省三亚市榆海路12号
Telephone : +86 (898) 8820 9999
Fax : +86 (898) 8820 9393
E-mail : mosan-reservations@mohg.com
Http : //www.mandarinoriental.com.cn/sanya/

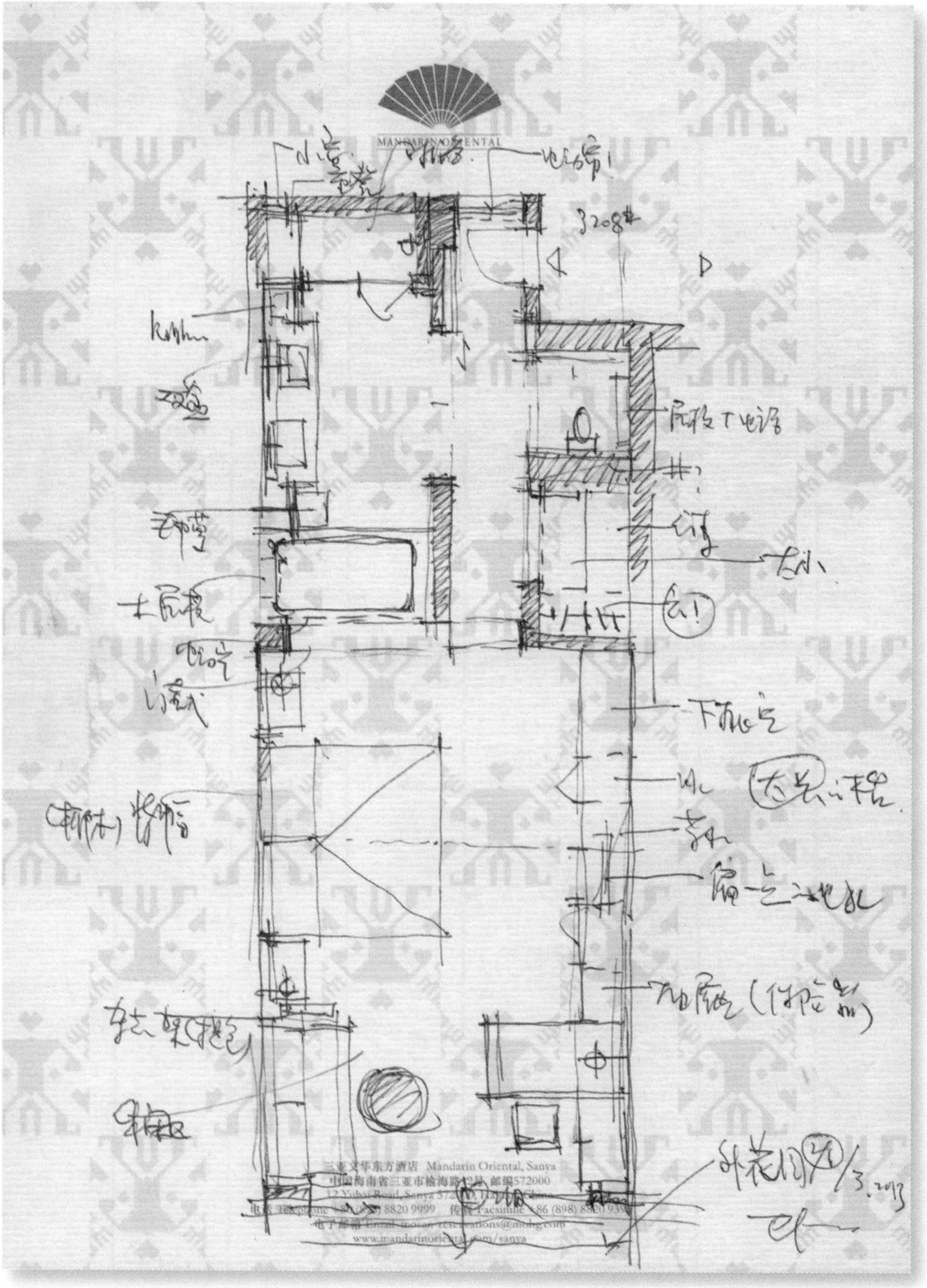

未更新已超换代

按惯例，6、7年要翻新一次之五星级酒店，或者因为竞争环境之恶劣，又或特殊气候，给酒店之消费人群造成损坏，而更应当缩短一些，不然会“惨不忍睹”！

三亚项目出差，与美女同事一起，为方便便住回几年前开业之位于大东海之文华东方酒店，而且性价比不错！

美女们二层之房间，我住六层之！可惜再进入房间，一股扑面之发霉味道，墙面陈旧，洗手间选用浮色之石材，经几年之洗冷用后之浸蚀，发旧，很不干净之印象，一切一切都显旧了。才几年之工夫，要保养真不容易。我相信酒店方面已经花了大力气了，但无耐使用，气候等因素让室内之墙层走向“老去”，再推开越的到达好像很久没人到访之旧阳台，心再次下沉了一下，这可是几年前之标杆酒店啊！近年同档酒店如雨后春笋，不容置疑之开房率会大大打折（春节旺季除外）这种状况如不及时改变，或翻新或大手笔改造，加上重新营销，那生意还是很成问题的！

酒店，特别是在三亚的竞争长期的老化，残酷而持续，
或许，真是未及更新，已要换代了.

Upgrade before You can Realize It

未更新,已然换代

按惯例六、七年翻新一次的五星级行规或者因为竞争环境恶劣，又或特殊气候，高频的消费人群造成损坏而应当缩短一些，不然会“惨不忍睹”！

In accordance with the customary convention, renovating one time every 6 or 7 years is the industry rule for the Five Star Hotel. Due to the bad competing environment or special climate and high frequency of consuming crowd, the renovation period will be reduced and ahead of time. Otherwise it will be ruined!

三亚项目出差，与美女同事一起，求方便便住回几年前开业的位于大东海的文华东方酒店，而且性价比不错！

When I went on an errand with my beauty colleague for one Sanya Project, we lived in Mandarin Oriental Hotel located in Dadonghai for the sake of convenience, which was opened several years ago and the quality versus price performance was also very good!

美女住了二层的房间；我住有后花园的！可惜再进入房间，有一股扑面的发霉味道，墙面陈旧，洗手间选用深色的石材，经几年的洗涤用品的侵蚀，已发白，给人很不干净的印象。一切一切都显旧了，才几年的工夫，要保养可真不容易。我相信酒店方面已然花了大力气了。但无奈使用、气候等因素让室内的尊容过早地“老”去，再推开趟门到达好像很久没人到访的花园平台，心再次沉下去了一下。这可是几年前的标杆啊！近年同档酒店如雨后春笋，不容置疑的开房率会大大打折（春节旺季除外）。这种状况如不及时改变，或翻新或大手笔改造，加上重新营销，那生意还是挺成问题的！

三亚文华东方酒店 *Mandarin Oriental, Sanya*

Our beauty colleagues lived in the room in the second floor. I lived the room with back garden! When I went into the room, one moldy smell blowed against my face. the wall was old and stone plate with strong color was applied on the washing room. The washing items which after several years of corroding had became white and dirty to my first impression. All became too old. It was just several years and actually keep maintenance is never easy. I believed that the hotel should take great pain to make decoration. However, no use as climate factor made the interior decoration decayed, then pushed open the door and arrived in the garden platform which seemed no one visited. My heart became so heavy and originally this was the landmark several years ago! During the recent years, the hotel sprang up and the room opening rate became sharply decreased (except the peak season during the Spring Festival). If you did not change timely or making new renovation or re-adjust the marketing strategy, then business became the touchy problem!

酒店，特别是在三亚的竞争长期白热化，残酷而持续。或许，真是未及更新，已要换代了。

Hotel, especially the long-term fierce competition in Sanya, cruel and continuous. Maybe, it is time for the hotel service to be upgraded before you can realize it.

上海外滩悦榕庄酒店

22 ★★★★★

BANYAN TREE,SHANGHAI ON THE BUND

Address : No.19 Gong Ping Road, Hong Kou District Shanghai 200082, People's Republic of China.
中国上海市虹口区公平路19号

Telephone : +(86) 21 25091188

Fax : +(86) 21 25092288

E-mail : shanghaionthebund@banyantree.com

Http : //www.Banyantree.com/

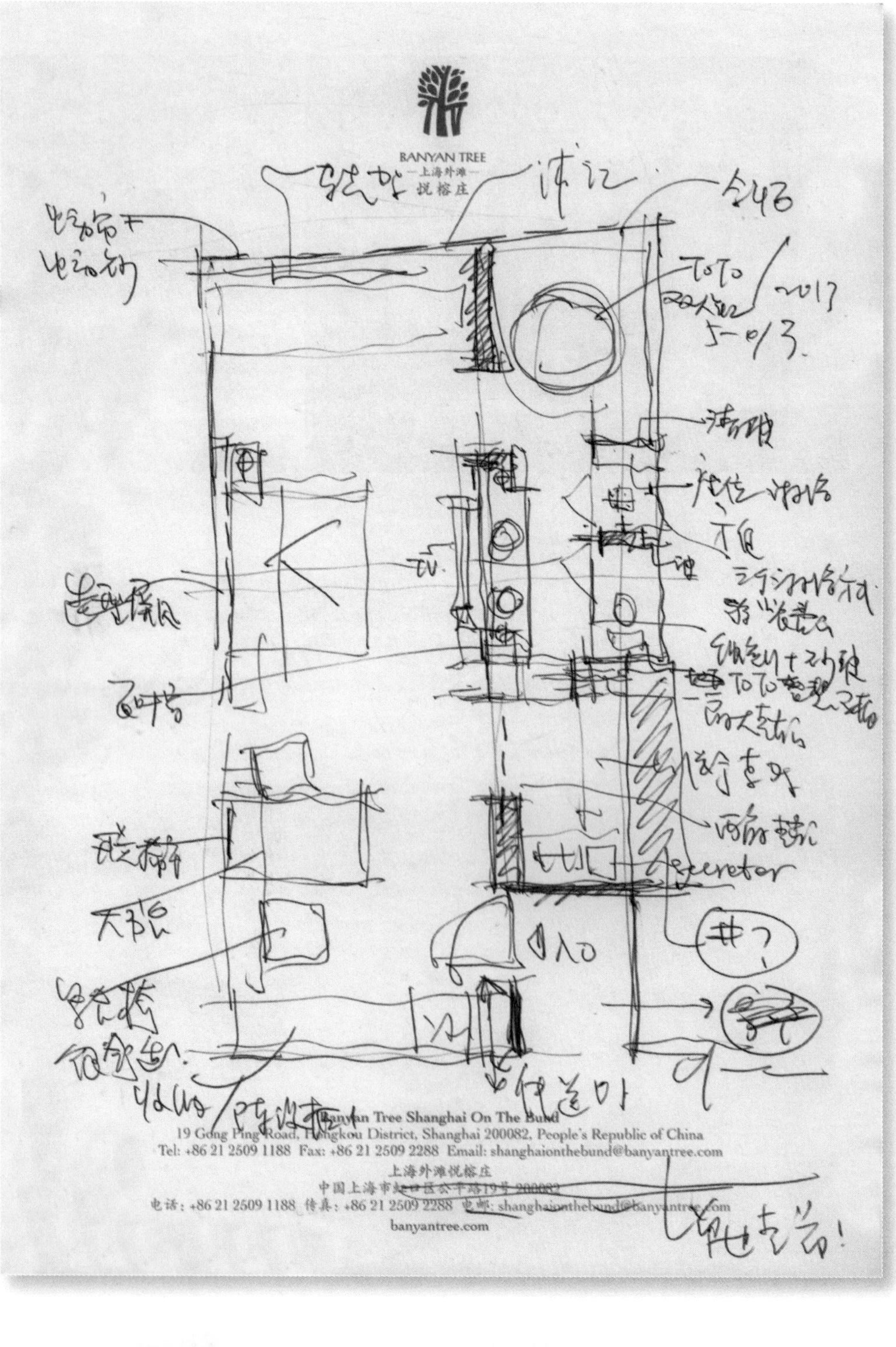

BANYAN TREE
上海外滩
悦榕庄
Banyan Tree Shanghai On The Bund
19 Gong Ping Road, Hongkou District, Shanghai 200082, People's Republic of China
Tel: +86 21 2509 1188 Fax: +86 21 2509 2288 Email: shanghaionthebund@banyantree.com
上海外滩悦榕庄
中国上海市虹口区公平路19号 200082
电话：+86 21 2509 1188 传真：+86 21 2509 2288 电邮：shanghaionthebund@banyantree.com
banyantree.com

"两不误"

听闻度假品牌 Banyan Tree 悦榕庄要在上海了，还在外滩路段，没得说又是惊喜。地处外滩哪里还有这等条件让这个品牌打造"世外桃园"。

打的(Taxi)，司机总是找不到，一路问，后来，很难找的路段，包括入口景观、雨篷形式，完全不是我印象中的那个"朴素"女子了。进门大堂开始润泽，装置，走道，公共区域，结束入房间……

还好，房内还是有那种味的感觉，简单的设计，木色微沉，灯光浑圆，细节到位，当然还有大大的双人大圆浴缸，静宓，浪漫沛沛，让你有宽衣解带的冲动！

一个以度假休闲为本的酒店，入驻到商务氛围十分的上海外滩，现代的外表包裹传统的东方亚洲，也可谓一种新的"招式"住下来还是感觉能友好地融合在一起，当然这样可以兼顾如我这样喜欢 Banyan Tree 品牌的时尚商务人士；也可以今后大都市时尚有加的大富贵，可谓品牌推广与开拓市场的"西不浅"

当然从大平面布局，单也走廊，全面的布置空间，再到房间的细节考虑到业主心思的，大空间的分隔与划分，一步到位：主次、动静；功能与情趣、细节与舒适都满足了；小区域细致到位，目不暇接，大大的储物柜，精心收纳，整体框图合成墙，大大的客厅区，可以让孩子尽情的放下大专行李箱的衣帽间，环岛式的双盆洗脸区，梳妆区，以及精致私密的沙发等（每套都各不相同）当然还有可以提到的，情人的双人圆形浴缸，一切都值得回味，留下难以磨灭的印记！

可以说，是这期开发中最值得推荐的设计面之一，不知道已住进的你有没有同感！？

欢迎"西不浅"，学习和借鉴，欢迎找我！

Both compatible
"两不误"

听闻度假品牌Banyan Tree悦榕庄来上海了，还在外滩路段，满是欢喜。想想外滩哪里还有这等条件，让这个品牌打造“世外桃源”。

I heard that the holiday brand Banyan Tree came to Shanghai. More important that it is still located at the WaiTan Section and I feel quite happy. We just thought about which place can have such good condition like "land of idyllic beauty" that helps build up this brand.

打的（Taxi），司机绕了几圈才艰难找到。一看，啊，很现代的外观，包括入口景观，雨篷的方式，完全不是我印象中的那个“朴素”女子了。进了大堂开始调整，装置、走道、公共区域，待进入房间……

Took a Taxi and the driver made several turnarounds and I finally got it. Just had a look at, well, it was one hotel with quite modern outer appearance including the entrance landscape, way of raining awning, completely different from that "naïve and simple" country girl that left in my impression. Entering into the lobby, you will feel quite different, starting from the device, the aisle, public area and then entering into the room, you will never be deceived by its too modern outer appearance, instead it is quite simple, traditional and countryside.

还好，房内还是有原汁原味的感觉，偏东方的设计，木色微沉，灯光浑圆，细节到位，当然还有大大的双人大圆浴缸。靠窗，望黄浦江，让你有“宽衣解带”的冲动！

Fortunately, the room inside still keep the original sense and original taste, a bit oriental design, slightly heavy with the timber color,and light is strong and even, the details are excellent. Of course, there are round bath tub for two people. Lying on the window, overlooking the Huangpu River, you may have impetus for taking off clothes and go into the river for swimming!

上海外滩悦榕庄酒店 *Banyan Tree , Shanghai on the Bund*

一个以度假休闲为本的酒店，入驻到商务气氛十分浓的上海外滩，现代的外表包裹传统的东南亚调子，可谓一种新的“招式”。住下来还是感觉能友好地融合在一起，当然这样可以兼顾如我这样喜欢*Banyan Tree*品牌休闲特色的商务人士，也可以分享大都市时尚市场的大蛋糕，可谓品牌推广与开拓市场“两不误”。

One hotel that taking the holiday leisure as the basis, permanently stationing in the Shanghai WaiTan, traditional ASEAN tone is wrapped by the modern outlook, This is one new "trick" to attract the customers as well. You will feel that you can be friendly merged with it together when you live there. Of course, this can help take care of the business person who loved Banyan Tree leisure brand. Banyan Tree also shared a big cake in the metropolitan fashionable housing market. Both brand promotion and market development are taken care of.

当然从大平面布局、单边走廊、全面江布置房间，再到房间内的细节都是煞费心思的。大空间简单划分，一步到位。主次、动静，功能与情趣，休闲与商务都满足了。小区域细致到位，目不暇接：大大的陈列柜，整体收纳；茶水柜组合成墙；大大的写字区；可以让你长驻的，放下大号行李箱的衣帽间；环岛式的双盘洗脸区，梳妆区；江景倚窗看书的沙发床（可随手拿到杂志）。当然还有前面提到的撩人的双人圆形浴缸，一切都保留有品牌的深度和印记！

Of course, lay out from the big plane with single corridor,from the room you can see the comprehensive river view, the details inside the room are so particular. Big space is simply divided and one step to reach. Primary and secondary, movable and static, functional and sentiment, leisure and business, all of them you can find here. For the small area, it is detailed and particular. You will see varieties of cabinets including big display cabinet, storage tank and tea cabinet that are formed into the wall, big writing area, you can permanently stay there, the big clothing room where the big size of luggage can be placed, island type of dual-plate face-washing area, grooming area, river-view book-reading sofa (magazines are all available). Of course, there are round-shape bathing tub for two-people that mentioned above. All keep the depth and imprint of the original brand!

可以说， 这是最近开业中最值得推荐的江景酒店之一。不知道已住过的你有没有同感？！

I can say that it is one of the Riverview Hotels that open recently and most worthy of being recommended. I don't know if you share my feeling?

确是“两不误”，学习和休闲，对于我！

For me, study and leisure both can be perfectly compatible.

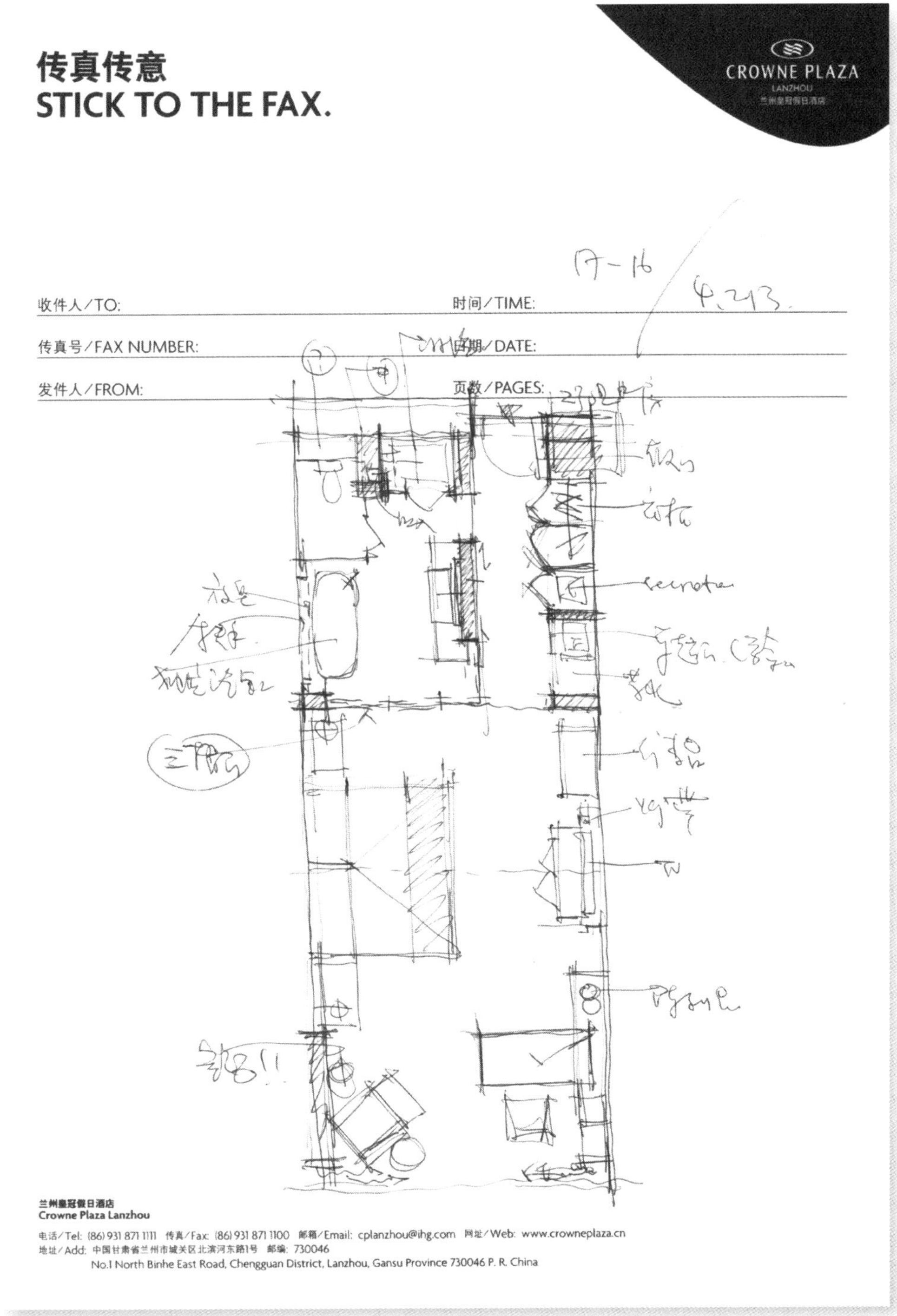

传真传意
STICK TO THE FAX.
CROWNE PLAZA
LANZHOU
兰州皇冠假日酒店
收件人/TO:
时间/TIME:
传真号/FAX NUMBER:
日期/DATE:
发件人/FROM:
页数/PAGES:
secretar
兰州皇冠假日酒店
Crowne Plaza Lanzhou
电话/Tel: (86) 931 871 1111 传真/Fax: (86) 931 871 1100 邮箱/Email: cplanzhou@ihg.com 网址/Web: www.crowneplaza.cn
地址/Add: 中国甘肃省兰州市城关区北滨河东路1号 邮编: 730046
No.1 North Binhe East Road, Chengguan District, Lanzhou, Gansu Province 730046 P. R. China

CROWNE PLAZA
LANZHOU
兰州皇冠假日酒店

兰州皇冠假日酒店

23 ★★★★★

CROWNE PLAZA, LANZHOU

Address : No.1 North Binhe East Road, Chengguan District,lanzhou, Gansu Province 730046 P.R.China
中国甘肃省兰州市城关区
北滨河东路1号

Telephone : +86 931 871 1111

Fax : +86 931 871 1100

E-mail : cplanzhou@ihg.com

Http : //www.crowneplaza.cn/

谁"最怪"!?

长方盒子能折腾出什么来，对于酒店客房，优选的长方盒子基本形，这是我从这几年住的酒店，以及经历其发展的趋向，当然是以较熟悉的来作例，建筑本身古灵精怪的，暂不包含，这已够我头痛的了！

可以说酒店的客房布局是越来越怪了，"怪"的有两个方面，一是水平不行而"创意"之怪，二是有水平而突破的。

第一种因无价值而不讨论，我们来看看第二种，也分为认同的和不怎么认同的。

我们不妨看一看北京的柏悦酒店，超长的平面，与几年前其君悦自身品牌的王府井君悦酒店平面类同（请见第一本《住哪》P134）而且两个酒店斜对门，相距不过四百米，或者各个品牌也固化了自己的模式，只是我们作为入住的人们也不甚满意而已！

倒是随着客房的开间越来越宽，从4.2米到4.5米、4.8米甚至更宽，布局也从传统的单边式、发展到"回归"的走廊左右布置功能区域，而且变化也越来越多，或有嵌入式的小衣帽间，如长沙万达文华酒店；或丽浴分离，如深圳东海朗廷酒店，兰州皇冠假日酒店，小变化，大雷同，更有甚者，两者兼有。今后更加优化的，如三亚文华东方酒店，走道一边是两个衣帽间，而另一边是洗浴区与淋浴浴缸区域，当然配以多样半开放、全开放的布局；半透明、透明的亦有，但宽度是最重要的条件。总之客房的"套路"手段随着客房面积的不断升级而越演越烈，于是乎求新，客房的"怪"招也就层出不穷了！

"怪"有时候会让人遐想，让短住的你我有了特殊的喜悦和含义，也揣摩了设计师的思想，当然亲身体验，才是最重要的！

作为设计师的我，住多了，也慢慢地觉"怪"不怪了！

Who is the most "Strange"?

谁最“怪”！？

长方盒子里能折腾出什么来？对于酒店客房，就是以长方盒为基本形。这里我就这几年住的酒店，回头看看其发展的趋向，当然只是以较规矩的来作例。建筑本身古灵精怪的，暂不包含，这已够我头痛的了！

What idea or artistic works can be made from one rectangle box shape of hotel? For the guestroom at the hotel, they took the rectangle box as the basic type. Here I'd like to share with you my experience when I lived in this hotel these years and looked back upon the trend it will develop. Of course, we just take the normal one as an example; architecture itself is quite unique and strange. We just neglect it as this is already a headache for me.

可以说酒店的客房布局是越来越怪了。“怪”有两种方向：一是水平不行而“创意”之怪；二是有水平而突破的。

We can only say that the layout of the guestroom of the hotel became more and more strange. There are two kinds of directions for "strange".The first one is strange but creative based on poor skill with less of art. Another one is unique strange and creative based on good skill with full of art.

第一种因无价值而不讨论。我们来看看第二种，也分为认同的和不怎么认同的。

We do not want to comment on the first one as it no value and not worthy of comment. Let's look at the latter that is creative and innovative based on good skill. This kind of innovation is also divided into highly-recognition and non-recognition.

我们不妨看一看北京华尔道夫酒店，超长的平面，与几年前其希尔顿自身品牌王府井希尔顿酒店平面类同（详见《住哪?》P134）。而且两间酒店斜对门，相距不过四百米。或许各个品牌固化了自己的定式，只是我们作为入住的人们也不甚注意而已！

Let's have a look at the Waldorf Astoria Hotel in Beijing. It is one super-long plane in layout. It was similar to the Hilton brand in early period several years ago and also similar to the plane and layout of Wangfujing Hilton Hotel(Details please find my first book "Where to live" P134). Moreover, the two hotels' door is opposite and only four hundred meters away. Actually each brand has its own definition of art and form. However, we never notice that when we lived in the hotel before.

倒是随着客房的开间越来越宽，从4.2米到4.5米、4.8米甚至更宽，布局也从洗手间单边靠，发展到“回归”的走廊左右都有功能区域。而且变化也越来越多，或有步入式的小衣帽间，如长沙万达文华东方酒店；或厕浴分离，如深圳东海朗廷酒店、兰州皇冠假日酒店，小变化，大雷同；更有甚者，两者兼有，分得更加彻底的，如三亚文华东方酒店，走道一边是厕所加衣帽间，而另一边是洗脸区与淋浴浴缸区域。当然形式多样，半开放、全开放的亦有，半透明，透明的亦有，但宽度是最重要的条件。总之客房的“奢华”程度随着单房面积的不断升高而越演越烈，于是乎求新、求变的“怪”招也频繁出现了！

With the opening of guestroom became wider and wider, from 4.2 meter to 4.5 meter, 4.8 meter and even wider, the layout is also developed from the single-lying type of washing room into "returned" corridor with left and right functional area and the change became more and more. Some hotel offers small clothes room like Wanda Mandarin Oriental Hotel in Changsha. Another type is characteristic by separating between the toilet and the shower like The Langham in Shenzhen and Crown Plaza in Lanzhou, small change but all keeps the same. Even some have the both and separate it more thoroughly like Mandarin Oriental in Sanya, at one side of the corridor is the toilet and clothes room. Another side is for washing room and shower area. Of course they are available in various forms, semi-open, all-open, semi-transparent and transparent, however, width is the most important condition. All in all, the luxurious extent of the guestroom will be more fiercely extravagant with the rising area of the single room, then we need to make innovation and the so-called new trick for change is appearing more and more frequently!

“怪”，有时间会让人遐想，让短住的你我有了探求的喜悦和欲望，也揣摩设计师的思想。当然亲身体验，才是最重要的！

"Strange and Unique" gave us the time for thinking which let the customer who is living here only for a short time have the pleasure and curiosity for exploration, also study the thinking and idea of the designer. Of course, when you can experience by your own, that is the most important thing!

作为设计师的我，住多了，也慢慢地见“怪”不怪了！

Me, as the designer, living for so many times, slowly and slowly, I feel nothing special now.

兰州皇冠假日酒店 *Crowne Plaza , Lanzhou*

抒写此刻
TAKE A MOMENT...

CROWNE PLAZA
LANZHOU
兰州皇冠假日酒店

①

17-16/4, 2013. 入住兰州·皇冠假日酒店
住十一层 CROWNE PLAZA.(LANZHOU)

再入住的感受

这家半年前曾住过一次，和同事、本地好友。但不能完成不到更好的当地酒店的体验，于是这次"再"入住。

半年后在此再住，感觉半年前都知道的酒店，入住就是一种熟悉的方式。虽然已经有了感情，正如中国酒店也开始认识到 Five Stars Hotel 一样，对于用户来说，酒店管理的设计、标准、服务等等。相信自己对待感情的方面去做，而作为酒店生意的人似乎更难些。正如同样的(酒店)二次入住，而我去看我的感觉(和从前)与酒店管理的(包括~~世界~~服务)的"新婚"了。相信以后更是亲密！现在嘛，在体验中不断更新！

喜好各个酒店有意义者，自己有意更有感受者。不同的"寻觅"和对于各类世界的尊重、浪漫之感，更加浪漫

兰州皇冠假日酒店
Crowne Plaza Lanzhou
电话/Tel: (86) 931 871 1111 传真/Fax: (86) 931 871 1100 邮箱/Email: cplanzhou@ihg.com 网址/Web: www.crowneplaza.cn
地址/Add: 中国甘肃省兰州市城关区北滨河东路1号 邮编: 730046
No.1 North Binhe East Road, Chengguan District, Lanzhou, Gansu Province 730046 P. R. China

在期盼的什么！Rums 的释色了，毕竟十分在世人 ②

酒店或会有一个商业的经营，这就正是如此的一样。认识

"作品"就是如此。的就是我们不是很高贵的意义（这就是以后的

商用，那么这就是我们所说的地位）但是要给我们很好。我们

我们也会成为被评论的一部分。

说是挺有趣，没准有一定的学问，也有价值的所在。

同样也只有人情，我们的知识已经开始。

我们说"游"也是努力！

再次/因为，作品也是，也是这样的说法

Rums

不用多想，Just take a moment!

王昀

Cl—

Feeling about Staying again
再入住的感受

之前来参观过一次，和同事。本无惊喜。但确实找不到更好的当地酒店去体验，于是这次"再"入住。

I have been here for visiting with colleague before. There was nothing surprise. But I couldn't find a better choice, so this time I stay here "again".

中午在中餐消费。只有体验才能知道细节。入住就是一种很好的方式，虽然已缺少了激情。正如中国遍地开花的五星级Five-Star Hotels一样，对于开发商、酒店管理方、设计方、施工方、供应商等，相信能保持激情的少而又少，而作为生意的似乎更现实一些。正因为生意（开房率）的不理想，近年有相当的发展商（投资方）与酒店管理方（包括世界级的）的"离婚案"。相信以后更是常态！观念嘛！在现实中不断更新！

I had lunch in the Chinese restaurant. Only experience can bring me detail information. Staying in hotel is a good way for experience, although I already lost passion. There are many five-star hotels all around China. For all the developers, hotel management agents, designers, constructors and suppliers, I think few people can keep the passion for hotel, but just only treat it as business. Recently quite a few developers (or investors) stopped cooperation with hotel management agents, including some global brands, just because the hotel occupancy is not good. I believe it won't be news in future! The ideas are always changing in reality!

幸好各个行业有新入者，有已存者，更有成功者。不同的"年龄"和对各专业的交替沉沦正好刺激着在其中的你我！所以就释然了。虽然十家开业的酒店或会有一两间有惊喜，但大家还是如生活一样，新的"作品"层出不穷，虽然自认为不见得是个好差事（关键未被认可，未成为这领域的专家），但只要坚持与参与，相信我们也会成为被评谈的一部分。

Luckily, there are new members, existing members and successful members for each industry. The different "ages" of members and the competition among different professions is stimulating us, who are working inside! It made me feel comforted. There are one or two hotels among ten new ones which can bring surprise, but new "works" still keeps coming out. Although I don't think it's easy for me to design a hotel (mainly because I haven't been regarded as an expert for hotel designer yet), I believe one day our works will become part of public topic, as long as we insist.

晚上拍拍照，是注重一间房的。也有体验的必要，用设计师心情。相信他们已然尽力。我们"到访者"也应尽力！

During I took photos at night, I concerned to one room, and tried to feel designer's mind. I think they already did their best, and we, as "visitors", should also do our best to feel!

用纸、用笔，做些消费，也正好是纸的抬头所说的。

Use the paper and pen to do something, just like the title of the paper.

不用多想，Just take a moment!

Don't think too much. Just take a moment!

上海衡山路十二号
豪华精选酒店

★★★★★

24

TWELVE AT HENGSHAN A LUXURY COLLECTION HOTEL, SHANGHAI

Address : No.12 Hengshan Road,Xuhui District
Shanghai,Shanghai200031China
中国上海市徐汇区衡山路12号
Telephone : +(86 21)3338 3888
Fax : +(86 21)3338 3999
Http : //www.starwoodhotels.com/

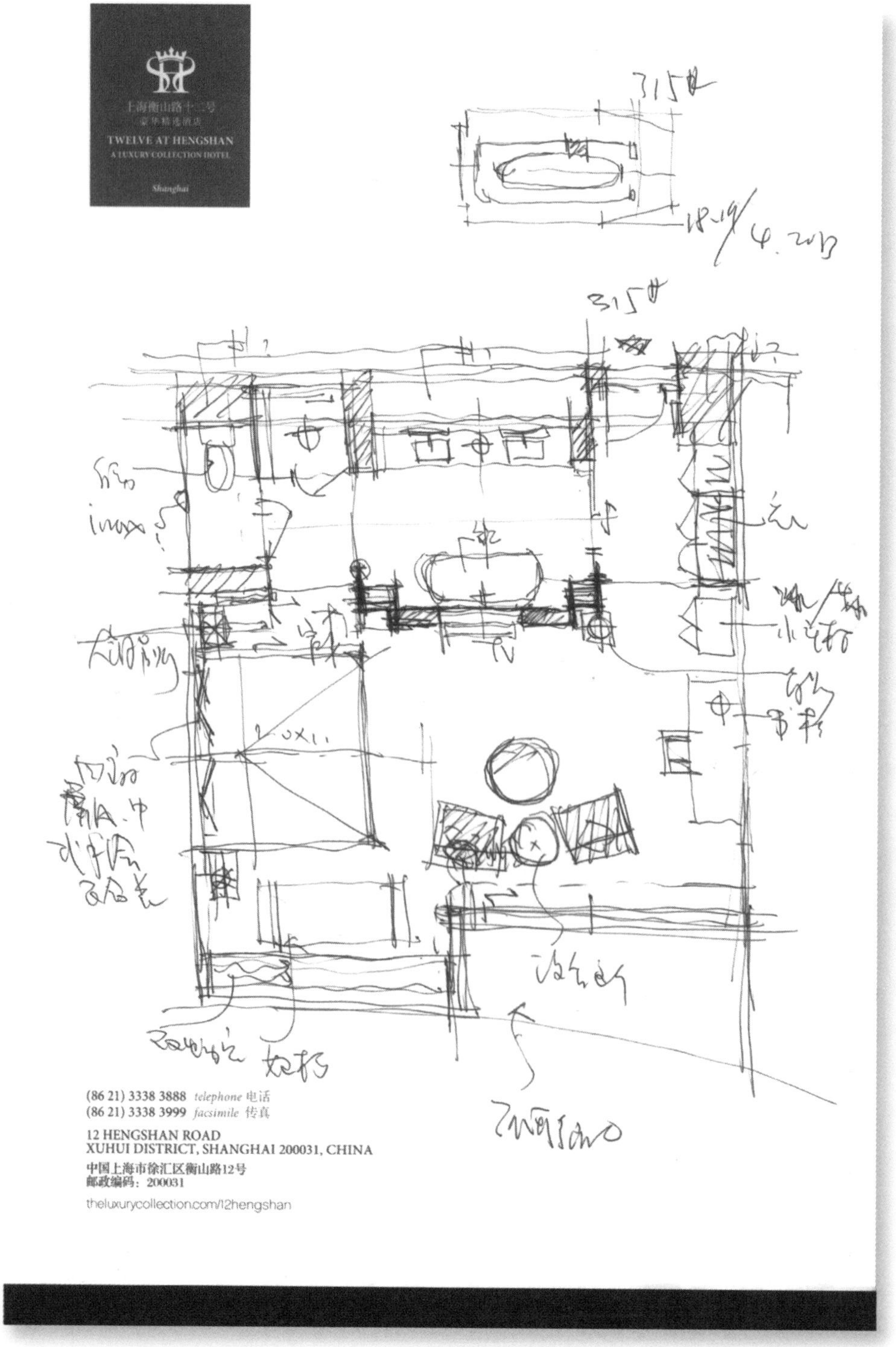
上海衡山路十二号
豪华精选酒店
TWELVE AT HENGSHAN
A LUXURY COLLECTION HOTEL
Shanghai
(86 21) 3338 3888 telephone 电话
(86 21) 3338 3999 facsimile 传真
12 HENGSHAN ROAD
XUHUI DISTRICT, SHANGHAI 200031, CHINA
中国上海市徐汇区衡山路12号
邮政编码：200031
theluxurycollection.com/12hengshan

双面信纸

住酒店多了，总爱找一些不同，双面信纸可谓偏少发现的。

这是刚刚开业的上海衡山路十二号豪华精选酒店，隶属喜达屋酒店集团，画出来用后，纸张，总觉得不一样，一翻背面，啊！信纸背面居然有句话："LIFE IS A COLLECTION OF EXPERIENCES. LET US BE YOUR GUIDE"（人生是各色经历的集锦，让我们带你体验！"双色的印刷，贴近酒店名号的"口号"，具煽动性和鼓励。

Yabu在中国的又一杰作，中国养了很多人，很多事，更加养了很多大师的作品。建筑是好"路子"，方形的外轮廓，与诸用地，斜切的进入式雨篷入口，和内凹形的内庭院，很中国，也很有国际范，是地传得到闹市中的静土！

高的楼层布局，让入住的客人在交互空间中走走，虽说只有几十米的门际，多数是不会打开窗帘，但细节可圈可点，彰显心思和功力。"轻松"的面貌也出色到位，包括画件册子。（大堂设计师）更合这种令人惊喜的信纸，中知道就对不住一下。双面写上画了。

之前也收集一些各酒店出品的信纸、便笺，其实也有几家是双面印刷的，如三亚的文华东方、美国科勒酒店。但多数只是浅色印底或单色，而不是让你使用的，但总比没有思考的到位、细心一些。看来，还是高端的更令人有宾至如归的感觉！

双面信纸只是一点，足以增加一点喜欢了。

Two-sided Letter
双面信纸

住店多了，总爱找一些不同。双面信纸可谓偶然发现的。

I had lived many hotels and I always liked to find something different. Two-sided letter was one discovery that found once in a while.

这是刚刚开业的上海衡山路十二号豪华精选酒店，隶属喜达屋酒店集团。画完平面后，看看，总觉得不一样。一翻背后，啊！信纸背面居然有内容“*LIFE IS A COLLECTION OF EXPERIENCES. LET US BE YOUR GUIDE*”（人生是各色经历的集锦，让我们带你体验）。双色的印刷，贴近酒店名号的“口号”，具煽情性和鼓励。

This is one newly-open hotel, where is located in No.12 of Hengshan Road, Shanghai. One hotel that is luxurious and extravagant, which is subordinated to Starwood Hotel Group. When I finished the plane layout and had a look at again and found it still different. When I turned it over and the back of the letter, one sentence aroused my curiosity, this letter paper had the words like "LIFE IS A COLLECTION OF EXPERIENCES. LET US BE YOUR GUIDE" written on the back. It was dual color printing and quite near to the slogan of the hotel, sensational and encouraging.

这是*Yabu*在中国的又一杰作。中国养了很多人，很多事，更是养了很多大师的作品。建筑是够“简单”，方形的外轮廓，占满用地，斜切的退入式雨篷入口，椭圆形的内庭院，很中国，也很有国际范，是难得的闹市中的净土！

It was another great works made by Yabu in China. Populous people are working here and many great works are produced from here as well from the great designer. The building is quite "easy", square contour, occupy full land and diagonally cutting and withdrawing type of awning entrance, oval shape of courtyard with quite Chinese characteristics, also it is quite international. It is indeed one quiet land that rarely to be found in the bustling city downtown.

上海衡山路十二号豪华精选酒店 *Twelve at Hengshan a Luxury Collection Hotel , Shanghai*

房间横向布局，让入住的客人有更开阔的景观。虽只有十几米的间隔，多数是不会打开窗帘，但细节可圈可点，颇有心思和功力。“软性”的配置也步步到位，包括宣传册子（大单张折页），更有这种令人惊喜的信纸。早知道就环保一下，双面写写画画。

The room is horizontal type in layout so as to let the guests have a broad view. Although it was only a dozen meter away, most of the curtains won't be opened. But the details were quite good and particular with art. "Software" configuration was also in place including the brochures (big sheet) and letter that makes you amazing. If I had known earlier, I would have drawn some pictures on two sides.

之前也收集一些各酒店品牌的信纸、便笺，其实也有几家是双面印刷的，如三亚的文华东方，美国科勒酒店，但多数只是满铺印纹，或单色，而不是让你使用的，但总比没有思考到位。细想一下，看来，还是高端的更令人有宾至如归的感觉！

Early on, I had also collected some letters and notes from some branded hotels. Actually there are some hotels making two-sided printing like Mandarin Oriental Sanya , U.S. Kohler. But most of them are only carved with printing thread or just one single color. It was for good looking and not for using. However, it was always good to consider this and make it carefully and the customers will find it high-end and feel comfortable just like staying at home!

双面信纸只是一点，足以增加一点喜欢了。

Two-sided letter is just a very little bit that make you like it, however, it is enough to capture its customer's favorite.

水

酒店房间里的水分三类，免送的瓶装饮用水；洗脸洗澡的冷热水，还有不被注意的小冰箱里的饮料酒水。

进门口渴，找水喝，不同的酒店有不同的配置，好点的有四瓶，洗手间两瓶，茶水柜台面或书桌上两瓶。当然了，经济些的可以无限量要求服务员送水给你喝。

瓶装水也体现酒店定位，多数的五星级酒店都是国产化，一则成本合理，二则支持国货。少有见到十分奢华的则免费提供进口的矿泉水，在以色列好像提的宾馆。感觉这个也变化了许多，前几年还能偶尔碰到免费的"依云"、"巴黎水"，现在呢？绝种了！

和冰箱里的比较一下，偶尔到新加坡或马尔代夫的度假村酒店会免费的饮用水、冰箱里的非酒精饮料，其实从个人角度也喝不了一两瓶，其他和酒都是会以口味和产品区分的，特别要说一下，打包！但是住多了酒店，常觉得小冰箱存放的饮料、酒类有些多余，或杞人忧天，甚至担心会不知不觉过期了！

住店，我常说最关心的两件事：第一个舒适性，当然包括床上用品：枕头、被子等等；第二件就是热水的稳定状况：水压和水量。现在的五星级酒店虽说是重金打造，双重保险，偶尔不管网加增压泵，但可能设计和结构的差异，印象中我住的酒店中能有绝对满意的不到一半，多数是水压不足，更有甚者是时冷时热，让你在淋浴过程中不断的跳前跳后，严重影响享受，有时候真想砸烂浴室！

宾馆的水，看来最重要和敏感的就是淋到身上的了！

下次你住店，不妨关注一下，可以灌一下小冰箱的饮料，用瓶装水煮煮咖啡，当然需要一个长一点时间的淋浴，验证一下水的状况，如果有耐心还可以放一缸水去泡泡，不过可需等上近十分钟啊！

Water
水

酒店房间里的水分为三类，配送的瓶装饮用水，洗脸、洗澡的冷热水，还有不被注意的小冰箱里的饮料（酒）水。

Water in the hotel room was categorized into 3:bottled drinking water that delivered, hot and cold water for washing or taking a shower, and the drinks (wine) water that stored in the refrigerator that left unattended.

进门，口渴，找水喝。不同的酒店有不同的配置，好一点的有四瓶：洗手间两瓶，茶水柜台面或床头柜两瓶。当然原则上你可以无限量地要求服务员送水给你喝。

When entering the door, you may be thirsty and need to find some water to drink, and different hotel has different setup. For the good one, there are 4 bottles available: 2 bottles for washing room, 2 bottles for the tea counter and bedroom. Of course, usually speaking, you can ask for more from the service woman.

瓶装水也体现酒店定位，多数的五星级酒店都是国产化，一则成本合理，二则支持国货。少有见到十分顶尖的品牌免费供应进口的矿泉水，担心这成为亏损的窗口。感觉这也变化了许多，前几年还能偶尔碰到免费的“依云”、“巴黎水”。现在嘛，绝种了！

Bottle water also showed the positioning of the hotel. Most of the Five-star hotels domestic. One is for its reasonable cost, the second is for supporting the national brand. However, seldom the top brand will supply imported mineral water for free of charge and worry this may cause loss. I felt that these years change quite a lot. Several years ago, we could still see the bottle water for free, such as "Yiyun", "Paris Water". Now all are gone!

和冰箱里的比较一下，遇到新开张的，或高端的度假精品酒店，会免费饮用小冰箱里的非酒精饮料。其实从个人角度也喝不了一两瓶，亦非所有的都是合口味和高品质的，除非环保一下，打包！倒是住多了酒店，常觉得小冰箱存放的饮料、酒类有些多余，或杞人忧天，老是担心会不知不觉过期了！

Compared with the goods in the refrigerator, when we meet the newly-open or high-end holiday hotel, we will drink the non-alcoholic drink in the small refrigerator for free. Actually from the personal stance, I don't think people can drink more than 2 bottles unless they are taken away. Not all the taste can be suitable for the customer.

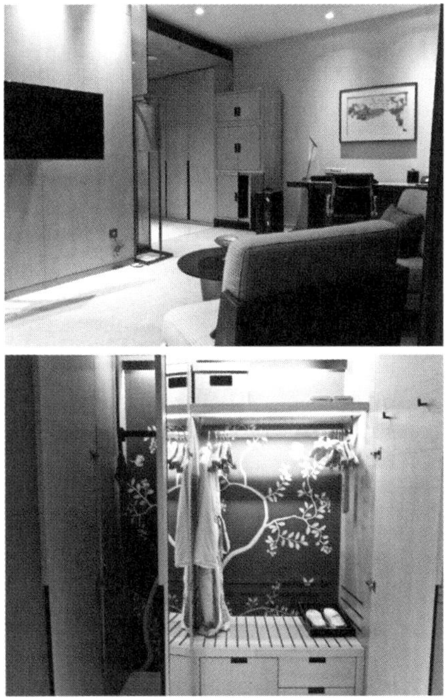

上海衡山路十二号豪华精选酒店 *Twelve at Hengshan a Luxury Collection Hotel, Shanghai*

Maybe it is considered from the purpose of environment-friendly. Pack them! Living more often in the hotel, sometimes I felt that the small refrigerator was filled with too much stuff. Actually some drink or wine is surplus and no use as they have the expiry date. I don't know if others will see it in my way or not and personally I am so worried about the expiry date if I really want to drink from the refrigerator!

住店，我常说最关心的有两件事。一件是床的舒适性。当然包括床上用品：枕头、被单等等。另一件就是热水的供应状况：水压和水量。现在的五星级酒店虽说是重金打造、双重保险，循环管网加增压泵，但可能是因为设计和经验的差异，印象中我住的酒店中能相对满意的不到一半。多数是水压不足，更有甚者是时冷时热，让你在淋浴过程中不断地蹦蹦跳跳，严重影响了享受。有时候真想砸烂浴室！

Talking about living in the hotel, I often talked about two issues that I concerned most. One is the comfort of the bed, including the bed items such us pillow, sheet, etc. Another one is the hot water supply condition, water pressure and water volume. Nowadays, although the Five-star hotel bluffed that their hotel are built by spending quite a lot of money and it is dual safe, pump with high-pressure in the pipe, etc. Maybe the discrepancy between the design and experience, from my impression and experience, less than half of the hotels can satisfy me in reality. Most of them are equipped with insufficient water pressure, even some are cool and some are hot water. You have to be hopping during the process of taking a shower and it severely affects the comfortable feeling. Sometimes I really want to break the bathroom as it almost drives me crazy!

房间里的水，看来最重要和敏感的就是淋到身上的了！

Talking about the water in the room, it seems that the most important and sensitive issue is checking the water supply facility and make sure taking a good shower and have a good time when staying in this hotel.

下次你住店，不妨关注一下。可以消费一下小冰箱的饮料，用瓶装水煮煮咖啡，当然要享受一个时间长一点的淋浴，验证一下水的状况。如果有耐心还可以放一缸水去泡泡，不过可要等上近十分钟啊！

Next time when you need to live in the hotel, you can concern more about this. You can consume the drinks in the small refrigerator, then use the bottle to cook some coffee and of course you need to enjoy a longer time taking a shower and check the washing facility. If you have more patience, you can make a tub of water to take a bath, however, it may take you about 10 minutes!

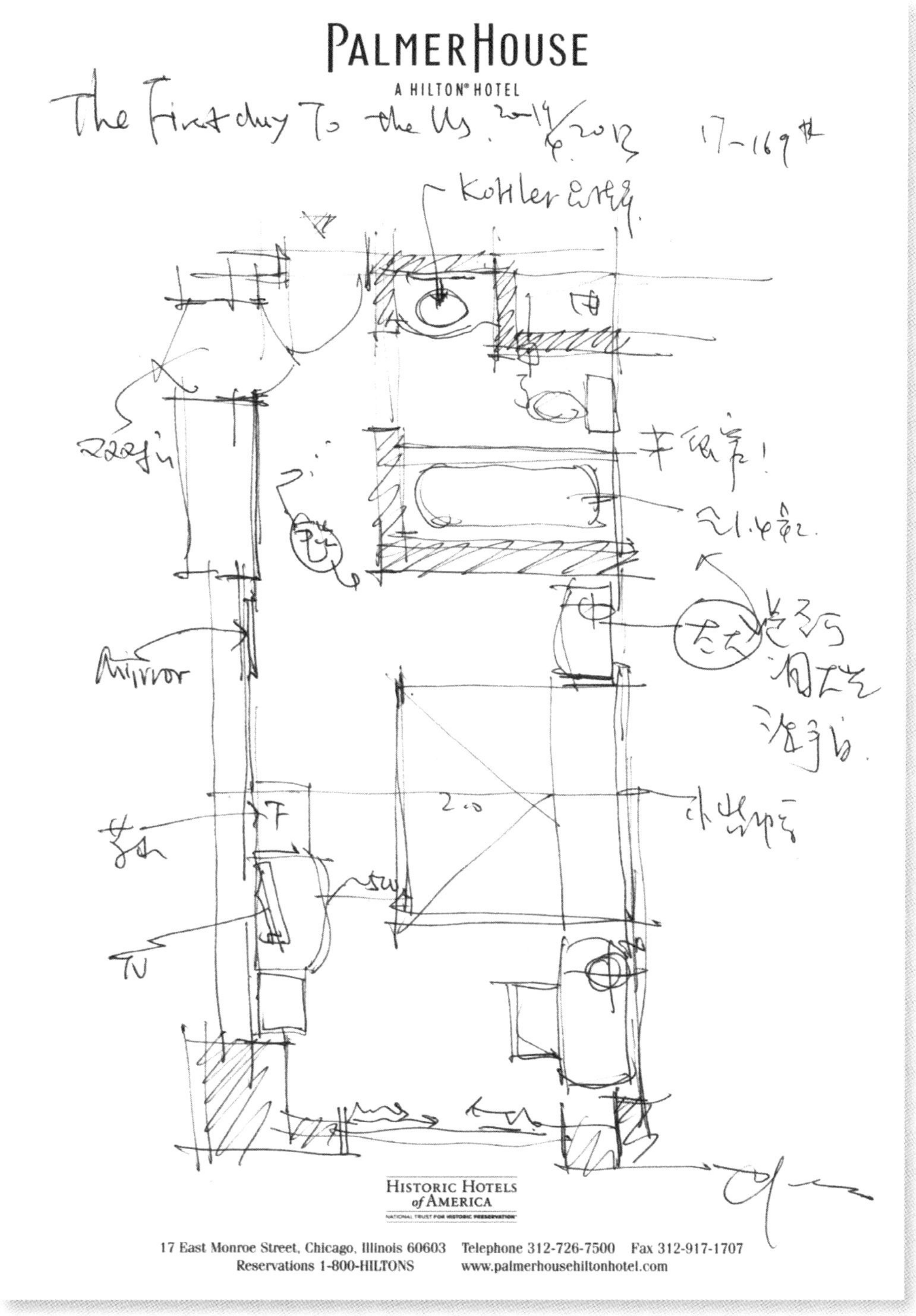
PALMERHOUSE
A HILTON® HOTEL
The First day To the US
Kohler
Mirror
TV
HISTORIC HOTELS of AMERICA
17 East Monroe Street, Chicago, Illinois 60603 Telephone 312-726-7500 Fax 312-917-1707
Reservations 1-800-HILTONS www.palmerhousehiltonhotel.com

PALMERHOUSE
A HILTON® HOTEL
希尔顿帕尔玛酒店

25 ★★★★

PALMER HOUSE, A HILTON HOTEL

Address : No.17 East Monroe Street, Chicago,Illinois 60603
Telephone : +312 726 7500
Fax : +312 917 1707
Http : //palmerhousehiltonhotel.com/

大酒店

什么叫大酒店？

我认为至少包含：一是总建筑面积够大，二是房间数够多，三是历史够长，帕尔马（palmer house）是希尔顿的品牌，之前没有住过也没听到，是我们这次美国之行中在芝加哥入住的点，前后共计入住了三次。可谓说是最深的酒店之一，1875年开业的"老人家"，一层有近百间房，十九层，（查了一下，共计有1639间房）是Historic Hotel of American成员之一。酒店大堂位于建筑物的二层，跨度大，古典，豪华，真可谓金碧辉煌，完全体现芝加哥在钢铁与革命的时代的先进性，同期我们还是"小家碧玉"的建筑阶段，可谓差距不少！

入住后，一团20多人，好不容易才各自找到自己的房间，太过瘾了，房间内部不像公共区域般豪华、繁琐，是简洁和舒适，更适合安静地休息，哈哈，看来还是挺聪明的，钱都花在了面子上，房间的浴室特别的小，有点不可思议，不知道大大的外国朋友

144

如何使用呢？

住了三趟，每次之房间位置不同，搞迷茫，甚至迷路几次，突然间觉得“大酒店”之名字可不是谁都能叫的。

这家艺术与奇之旅馆号称城中之城，确实见证着一个时期美国酒店业之辉煌！佩服！

希尔顿帕尔玛酒店 *Pabner House , A Hilton Hotel*

Grand Hotel
大酒店

什么叫大酒店?
What is Grand Hotel?

我认为至少包含：一是总建筑面积够大；二是房间数够多；三是历史够长。帕尔马（*Palmer House*）是希尔顿的品牌，之前没有住过，也很少听到。它是我们这次美国之行的中转站芝加哥入住点，间隔着分别入住了三次，可谓是"了解"最深的酒店之一。*1925*年开业的"老人家"，一层有近百间房，十几层（查了一下，共计有*1639*间房），是*Historic Hotel of American*的成员之一。酒店大堂位于建筑物的二层，跨度大、古典、豪华。真可谓金碧辉煌，充分体现芝加哥钢铁工业革命的时代领先性。同期我们还是"小家碧玉"的建筑阶段，可谓差距不少！

I think that at least the Grand Hotel should include these 3 points as below:1) The total construction area is large enough; 2) There are sufficient rooms to live; 3) History is long enough, such as Palmer House, which is the brand of Hilton. I seldom heard its name and had never lived in before. It was one transit in Chicago for this time of our American Trip. I lived here separately for 3 times and it was one hotel that I knew most. This hotel has been open since 1925. Of course it is old enough. Hundreds of rooms are available on the first floor, dozens of storey (I had checked, altogether 1639 rooms) and also is one of the member units of the Historic Hotel of American. The lobby of the hotel is situated on the 2nd floor of the architecture with big crossing, classic and luxurious. It is quite splendid and magnificent, fully embodies the leadership image of the Iron Steel Industry Revolution time in Chicago. In the same period of that time, we were still in the early stage of construction and far lag behind them!

入住后，一团20多人，好不容易才各自找到自己的房间，太迂回了。房间内部不像公共区域般豪华、繁琐，显得亲和和舒适，更适合安静地休息。哈哈，看来还是挺聪明的，钱都花在了面子上，房间的洗手间特别的小，让人有点不可忍受。不知道大大的外国朋友如何使用呢?

After moving in, a group team of 20 people, it takes us great pains to find each of our room as it is too roundabout. The room is not like public areas that are luxurious and tedious, affinity and comfortable. However, it is more suitable for sleep. Well, it seems that it is quite smart idea. As you can see, all the money is spent on the key and to the right point. The washing room is quite a small room that I can not stand it. I can't imagine how our fatty foreign friends can use it.

住了三趟，每次的房间位置不同，有点迷茫，甚至迷路了几次。突然间觉得“大酒店”的名字可不是谁都能叫的。

Having lived here for 3 times, the position of the room is also different every time. I am a bit confused or even lost my way for several times. Suddenly I found that "Grand Hotel" was the name not every one can have.

这家芝加哥的帕尔马希尔顿当之无愧，确能代表着一个时期美国酒店业的辉煌！佩服！

Of course this Palmer Hilton Hotel in Chicago deserved this name. It indeed represents the splendid age of the American Hotel Industry! I personally so admire!

Hilton
HOTELS & RESORTS

新奥尔良河畔希尔顿酒店

26 ★★★

HILTON HOTEL, NEW ORLEANS RIVERSIDE

Address : Two Poydras Street, New Orleans, Louisiana, 70130, USA
Telephone : 1-504-561-0500
Http : //www3.hilton.com/

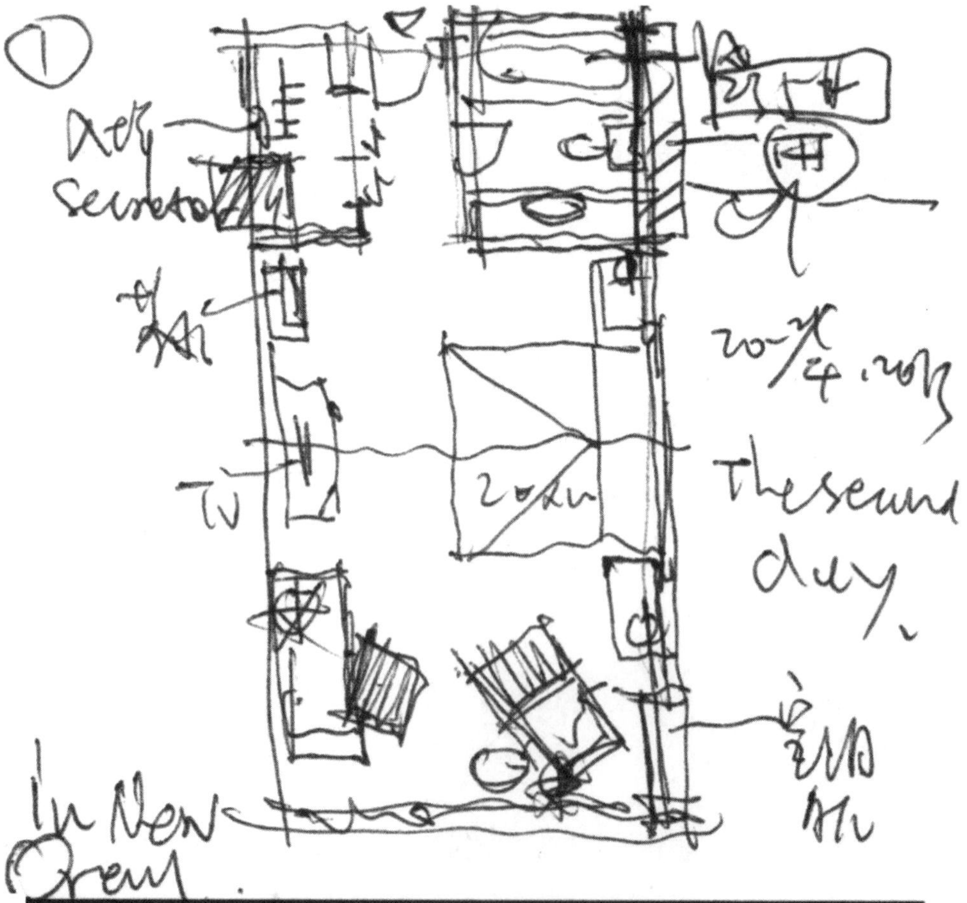

TV
The second day.
In New

AMERICAS · EUROPE · MIDDLE EAST · AFRICA · ASIA · AUSTRALASIA

Hilton
HOTELS & RESORTS
迈阿密市中心希尔顿酒店

27 ★★★

HILTON HOTEL, MIAMI DOWNTOWN

Address : No.1601 BISCAYNE BLVD, MIAMI, FLORIDA, 33132, USA
Telephone : +1 305 374 0000
Fax : +1 305 714 3811
Http : //www.hilton.com/

3/4th In Hilton (Downtown Miami)

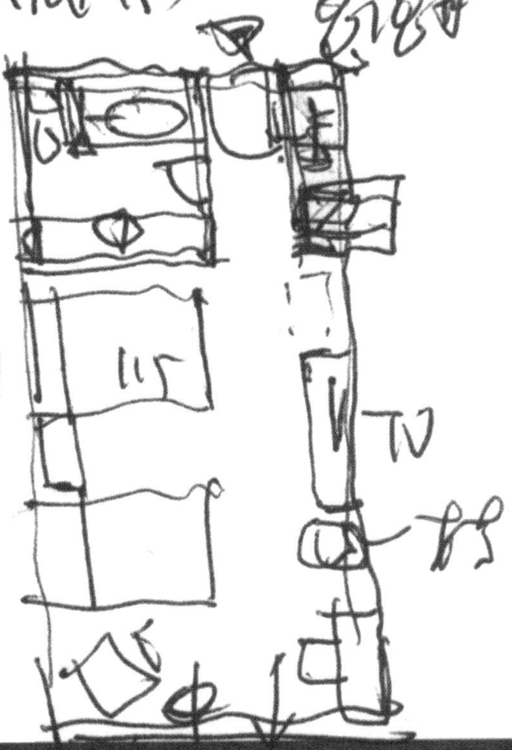

TV

AMERICAS · EUROPE · MIDDLE EAST · AFRICA · ASIA · AUSTRALASIA

The American club RESORT

A KOHLER EXPERIENCE

美国俱乐部酒店

28 ★★★★★

THE AMERICAN CLUB

Address : No.419 Highland Drive Kohler, WI, 53044

Telephone : +800 344 2838

Http : //www.americanclubresort.com/

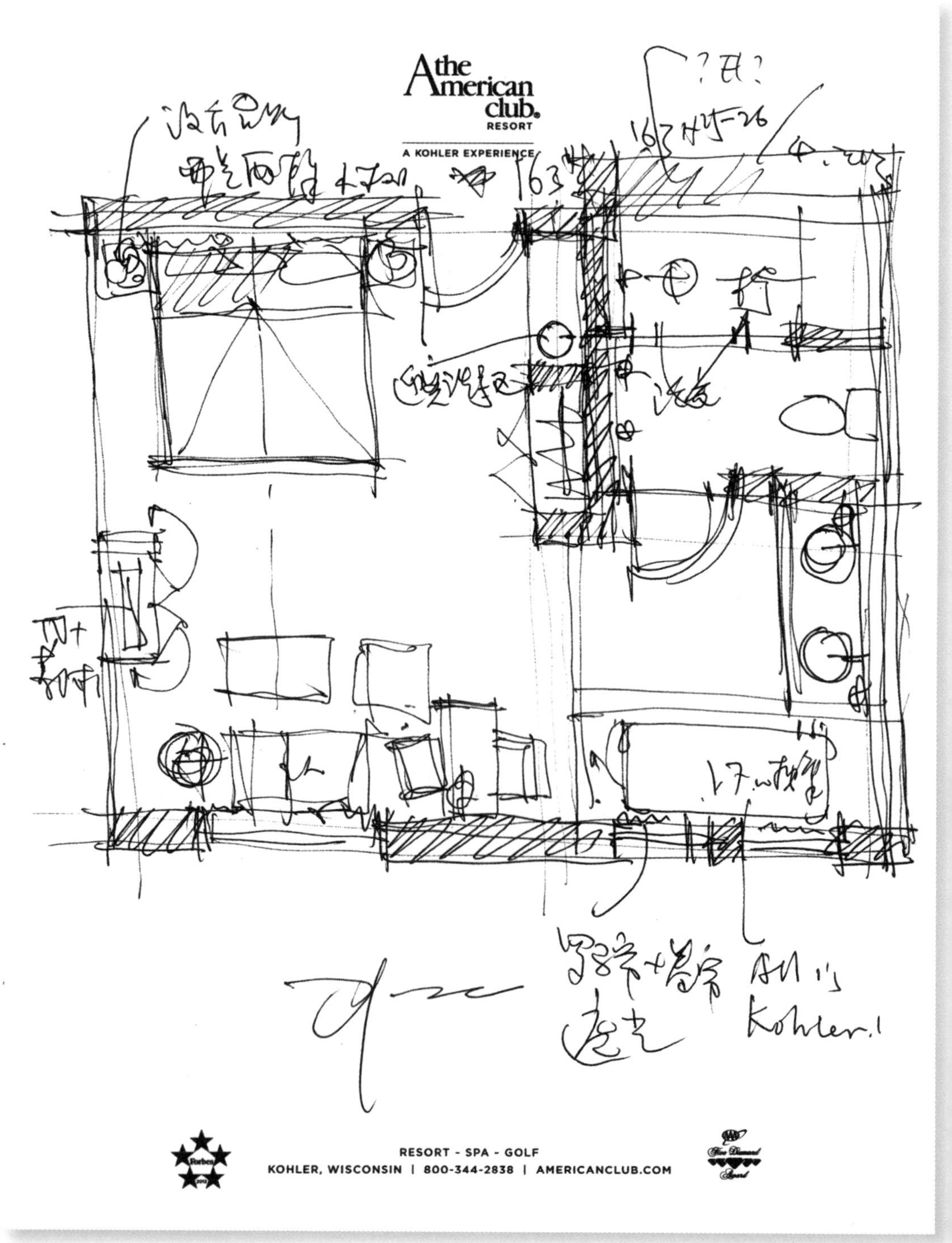

The American club RESORT
A KOHLER EXPERIENCE
RESORT - SPA - GOLF
KOHLER, WISCONSIN | 800-344-2838 | AMERICANCLUB.COM

科勒的酒店

你有没有想过住"科勒"品牌的酒店，其实你知道的事情真不少，不是吗？原来科勒经营着这么多、那么大的产业：旅游、地产、家具、酒店、艺术，当然还有更核心的洁具是大家最熟悉的一方。如今百多年的传承家族企业，中国还有这样的经典商人吗？

向你致敬，科勒！

科勒酒店，是由旧工人的宿舍改造而成，纯到不能再纯的大美风格，在蓝天、碧草、绿树下，经历着风吹雨打后的本色红墙，哈哈，真的是太喜欢、太漂亮了！

酒店的房间有什么特色呢？当然所有的卫生产品都是Kohler的：切割玻璃洗手盆、吊顶花洒，带喷射按摩头的淋浴间，经典款式的大浴缸，还有令我印象深刻的与众不同的地方，相信你一定猜不到。

房间入口处有一处巨大的洗手盆和化妆镜子，床头柜上没有床头灯，代替的是在床头柜上放了盆小梧桐，或者北美寒冷的天气，人们对绿色的渴望，全开放的洗手间，美式的直白与坦诚，洛杉矶向海岸的的花园，还有深夜时候可以慢慢享受园景，亦可看看电视。

细节更在于那精致门的五金件（五金合页的技巧），有机会我们也要好好引用到项目上。

住所的老经验，时尚用具，那本经典的秘诀，喝醉了的一夜，第二天一早起来慢慢地去享受了科勒的洗手间。

差一点忘了，全屋的家具更是科勒的品牌，世界级的 Baker（贝克）家具，舒服自然到极致！

这，就是科勒酒店！

Kohler Hotel
科勒的酒店

美国俱乐部酒店 *The American Club*

你有没有想过住“科勒”品牌的酒店？其实不知道的事情真不少，不是吗？原来科勒经营着这么多那么大的产业：旅游、地产、家具、酒店、艺术，当然还有支柱的洁具及配套的一切，140多年的传承家族企业，中国还有这样的经典老人吗？

Have you thought about staying in "Kohler" branded hotel? Actually there were so many things you did not know, weren't there? You never know Kohler run so much business and so huge industry, such as tourism, real estate, furniture, hotel, art, and of course it still ran its pillar industry like sanitary ware and its accessories. Inherited the family enterprise from more than 140 years old, I didn't think there was such a classic old man like him in China.

向你致敬，科勒！

I salute you, Kohler!

科勒酒店，是由以前工人的宿舍改造而成的，纯到不能再纯的大美风格。在蓝天、翠草、绿树下，衬托着风吹雨打后的灰红色外墙。哈哈，真的是太喜欢了，太漂亮了！

Kohler Hotel, it was transformed from the former staff dormitory. It was too much pure Pan-American style. Under the background of blue sky, green grassland and green tree, it made the grey red wall after the strong wind and heavy rain more outstanding. Aha. It is so fantastic, so beautiful scenery!

酒店的房间又有什么特色呢？当然所有的相关产品都是Kohler的：切割玻璃洗手盘，晶莹璀璨；带喷射按摩头的淋浴间；经典款式的大浴缸。还有令我印象深刻的，与众不同的地方，相信你一定猜不到。

What on earth is the characteristic of the hotel room? Of course all the related products are Kohler's: cutted glass washing plate, crystal and brilliant, The shower with massage spout and bath tub with classic style. Also there are some unique place that impress me. I believed you wouldn't got it.

房间入口处有一处迎宾的洗手盘和礼仪镜子，床头柜上没有床头灯，代替的是在床头柜上放了盒小松树。或许由于北美寒冷的天气，人们对绿色充满渴望。全开放的洗手间，彰显着美式的直白与坦诚。浴缸向酒店的内花园，泡澡时候可以慢慢享受园景，亦可看看电视。

There is one washing plate and etiquette mirror at the entrance of the room. No bed lamp is installed on the beside table, and one box of small pine instead of the bed lamp is placed there. Under the chilly weather in North America, people always admire the green color. All-open washing room shows American way of straightforward and sincerity. Bath tub is facing to the inner garden of the hotel, and you can enjoy the garden view while taking a bath or you can watch TV as well.

细节更在于淋浴间的防水性（五金合页的技巧）。有机会我们也要好好引用到项目上。

Details lie on the water-proof performance between the shower rooms (skill of the hardware hinge). We will try to apply them on our project if there is any opportunity.

传承的是经验，日积月累，那才是经典的秘诀。喝醉了的一夜，第二天一早起来慢慢地享受了科勒的洗手间。

Inheritance is the experience, accumulation from years to years, which is the classic secret. After a drunk night, enjoy Kohler's washing room slowly early the next morning.

差一点忘了，全屋的家具更是科勒的品牌，世界级的Baker（贝克）家具，舒服自然到极致！

I almost forget one point, that is all the furniture in the whole house is Kohler Brand, the world-class Baker Furniture, natural and comfortable.

这，就是科勒酒店！

This is Kohler Hotel!

THE PENINSULA
HOTELS

香港半岛酒店

★★★★★

29

THE PENINSULA HOTEL, HONGKONG

Address : The Peninsula Hong Kong
Salisbury Road, Kowloon
Hong Kong, SAR
香港特别行政区
九龙梳士巴利道

Telephone : 852 2920 2888

Http : //hongkong.peninsula.com/

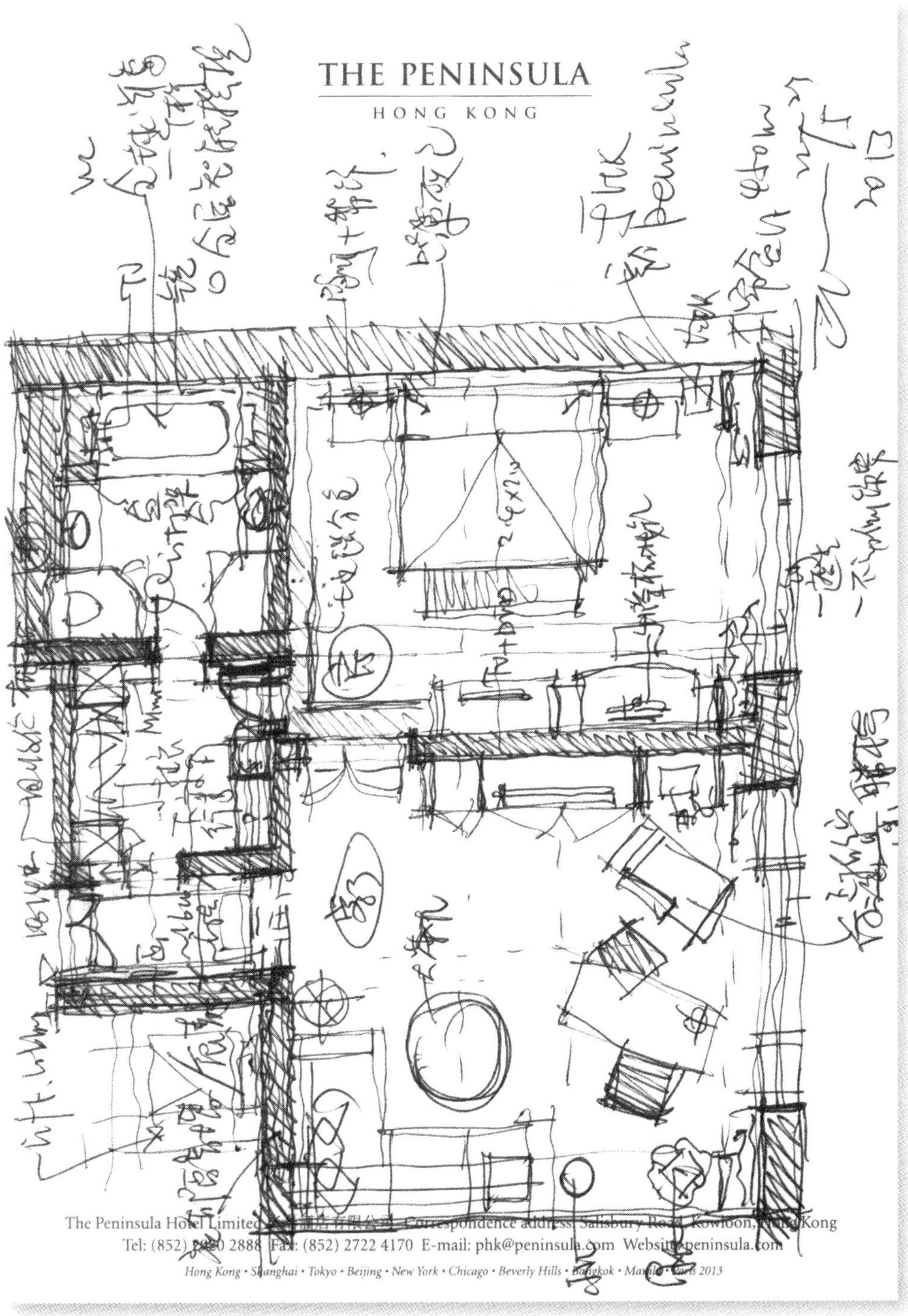
THE PENINSULA
HONG KONG
The Peninsula Hotel Limited
Tel: (852)
Fax: (852) 2722 4170 E-mail: phk@peninsula.com
Hong Kong • Shanghai • Tokyo • Beijing • New York • Chicago • Beverly Hills

城市桃源

有人说：看一个城市的经济发展指数，可以看看那里的银行大厦建得如何；可以看看市政府大楼的规模、豪华的宽敞尺度；那么如何看一个城市的生活状态，在广州可以看看喝早茶的人的笑脸；在成都看看街头麻将圈的围观人群多少；在北京看看奔驰宝马车的数量；在长沙夜间酒吧街的堵车指数，在上海看看外滩酒吧的俊男美女的数量……

不知不觉的，城市中的星级酒店也卷入了反映人们状况中了，我们更关注了，更融入了，也更亲近了，可以和朋友们喝喝下午茶，边吃边聊聊生意，中午吃个便餐，晚上泡个高级的酒吧，更有些特性的成为"旅游景点"，如广州的西塔四季酒店，空中大堂（70层）的人头涌涌，下午、晚上更成为观光景点，大堂吧常常爆满，北京的柏悦上海的金茂等等，更有如香港湾仔的英迪格酒店小酒吧君悦的"小城故事"，颇受欢迎。

特色的酒店可能会变成为去处，像个人士的会所。半岛的经典英式下午茶（其他酒店纷纷仿效：华尔道夫、四季、丽思卡尔顿）；半岛（上海）的长廊吧，四季酒店（杭州）的中午茶点及其间海鲜点心……不去住也可去享受其不同的区域，满足你不同时段和心情的需要，只要体会到。

平常没地方可去，不妨去走一走你身边的五星级酒店，相信一定会很满意、很relax，一个人或和你的朋友。

像连锁式的发展为会所、会员制，将酒店变为社会性，你更可将它当成你社交的家园，做做GYM、spa等等。普通一点的如和世度假村酒店、高档的也如安缦酒店般的。

城市的酒店，宛如成了一道风景，你身边的桃源，不在世外！

City Paradise
城市桃源

戏说：看一个城市的经济发展程度，可以看看那里的银行大厦建得如何；可以看看市政府大楼的规模，广场的宽敞尺度。那么如何看一个城市的生活状态？在广州可以看看喝早茶的人的笑脸；在成都看看街头麻将圈的围观人群多少；在北京看看景点黄包车的数量；在长沙看看夜间酒吧街的堵车程度；在上海看看外滩三里的俊男美女的数量……

香港半岛酒店 *The Peninsula Hotel, Hongkong*

One saying goes like this, If you want to see how fast one city develop, please just see how the construction of the bank building there or just see how large-sale of the building of the municipal government or how spacious size of the Square. Then the problem is how to see the living standard of a city? In Guangzhou, you can see the smiling face from the people who finish morning tea. In Chengdu, you can see how many people are crowded around the mahjong. In Beijing, please just see the rickshaws for scenic spot. In Changsha, please see the traffic jam at the night bar. In shanghai, please just count the number of the handsome guys and the pretty girls in Waitan Sanli.

不知不觉的，城市中的五星级酒店也“卷入”了反映人们状况中。我们更关注了，更熟悉了，也更亲近了。可以和朋友去喝喝下午茶，在堂吧聊聊商务，中午吃个便餐，晚上泡个高级的酒吧。更有地标性的成为“旅游景点”，如广州的西塔四季酒店，空中大堂（七十层）白天人头涌动，下午、晚上更成为观光景点。大堂吧常常爆满，比如北京的柏悦酒店、上海的金贸等等。更有如香港港岛的英迪格顶层小酒吧位于小城窄巷，颇受欢迎。

Unconsciously, the numbers of the Five-star Hotel in the city are also involved into the index for inflecting the living condition. So we are more concerned about that as we find it more familiar and more close to us. You can drink afternoon tea with your friends, talk about business in the lobby bar, have a fast food at noon, and stay in a luxurious bar at night. The city landmark becomes the "tourist scenic spot", like Air Lobby of Four Season in West Tower in Guangzhou (the 70th floor) Many people stay there in the daytime and it comes the sightseeing scenic spot in the afternoon and in the night. The lobby bar is quite crowded, like Park Hyatt Hotel Beijing, Golden Trade Shanghai, etc.. Even some are located in the shallow alley, like the small bars on the top layer of the indigo in Hongkong Island, and they are still quite popular.

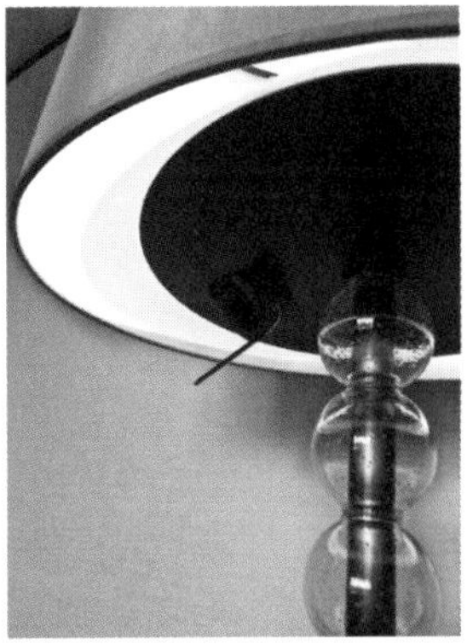

特色的酒店功能区域更成为商务、休闲人士的首选：半岛的经典英式下午茶（其他酒店纷纷仿效：华尔道夫、四季、丽思卡尔顿）；华尔道夫（上海）的超长吧台，四季酒店（广州）的中午茶市及其101海鲜餐厅……不去住也可去享受其不同的区域，满足你不同时段和心情的需要，只要你想得到。

The functional area of the characteristic hotel becomes the top priority for the businessman and leisure people. Afternoon tea of Peninsula Classic British Style (other hotel imitate and follow one by one: Waldorf, Four Season, Ritz- Carlton); the super-long bar in Waldorf (Shanghai) , Afternoon Tea in Four Season Hotels (Guangzhou) and 101 seafood restaurant... You can still enjoy any other area even you do not plan to live there. This hotel meets the need of different time section and the need of your mood, only if you can think of it.

午餐没地方去，不妨考虑一间你身边的五星级酒店，相信一定会很满意，很relax，无论一个人去还是和你的床友一起去。

If there is no good place for lunch, you can just consider the Five-star Hotel around you. I think you will be satisfied with it and will be relaxing no matter just alone or with your bedmates.

像连锁式的更发展为会所，会员制，将酒店资源社会化。你更可将它当成你放松的家园，做做GYM、SPA等等。普遍一点的如万达度假村酒店，高档的也如安缦酒店般的。

Many chain stores are even developed into Club or membership Club, socializing the hotel resource. You can take it to be your own relaxing garden, do the gym, spa, etc.. Ordinary one is like Wanda Resort Hotel and high-quality one is like Aman Resort Hotel.

城市的酒店，宛如成了一道风景。你我身边的桃园，不在世外！

The City's hotel is like beautiful scenery. It is the beautiful and relaxing garden that is never far away from you, and it is just around you and me!

Considerable for Each Detail
细节当道，“无微不至”

“无微不至”一般形容人物，但对于新翻新的香港半岛酒店来说，或者亦可！

Normally the word "considerable" is used for describing a person, but I think it is also suitable for Peninsula Hongkong, which is just reconstructed.

适逢酒店85周年之际，酒店各个区域分步进行彻底的翻新，从两三年前的部分客房开始到现在的望海景房间的全面升级完成，可谓耳目一新！

It's the 85th anniversary, and each area of the whole hotel was updated step by step, beginning from part of hotel rooms years ago, to the sea view rooms now. Everything is new!

这次吸取了上次未能入住的教训，特地提前一小段日子来订好大大的房间，一谓心诚之旅，如愿以偿。

Learned from the experience of missing booking, this time I made reservation for a big suite a couple days in advance, Finally I got what I wanted.

推门入内，硕大的玻璃窗正对维多利亚港。布局显得轻松随意，而如果深入使用，那可谓相当合理和细致，可谓让你佩服得不得了。

As long as pushing the door, there is a huge glass window facing Victoria Harbor. The layout of room seems to be easy, but if you stay and use, every detail is reasonable and thoughtful, which makes you admire very much.

一条环廻式通道连接小过廊式的入门，与行李房区、洗手间、客厅、卧室相通。客厅融合了休息、阅读、写字功能及电视一体柜、观海休闲沙发椅于一身，还有微斜布置的写字台、向窗斜靠的椅子。哈哈，你想不停留都不行，让你自然而然就躺下了。电动升降的插线板、沙发扶手的小杯托、落地灯下面的阅读小射灯、收藏式的电视、咖啡机，全屋智能灯光控制PAD，顶级、人性化。

There is a small hallway linking the entrance to luggage room, bathroom, living room and bedroom. In living room, the desk combines different functions of taking rest, reading, writing and TV. There are also a comfortable sofa for appreciating the sea, a slightly inclined writing desk, and a chair closed to window. Ha ha, you can't control to stay here and lay down unconsciously. The plug base can be raised and fell by power. A cup base is on sofa's armrest. A reading light is under the floor light. TV set and coffee maker can be hidden. All lights in the suite can be controlled by a pad. All things are high class and humanized.

香港半岛酒店 *The Peninsula , Hongkong*

入至卧室也有大大的高档电视、电动伸缩的梳妆镜，符合了中国人忌讳镜对床的传统。挂衣架与杂志架的结合，实用而有美感。关怀真谓无微不至。

There is also a big and high-class TV in bedroom. The dress mirror can be folded by power, so that the Chinese taboo of mirror facing the bed is solved. The clothes stand and magazine rack are combined, which is useful and pretty. The hotel is taking care of you in every detail.

还是老话，“软实力”无敌，正是半岛的杀手锏。当然改造不彻底也备受诟病，洗手间基本不变，与上述提到的区域格格不入。相信酒店方主要是担心噪声的影响太大。这着实让我看得不顺眼！

As I used to say, consideration for customer brings benefit. This is Peninsula's speciality. Of course there is also some disadvantage about partial reconstruction. The bathroom almost keep the same as before, so it looks so different from other area I mentioned above. I think that's because the hotel holder concerned about noise problem. But it still makes me uncomfortable!

纵观改造后的房间，确实让你流连忘返。大大的衣帽间，有专门的放鞋区、回收窗，以及可以满足你长期居住的挂衣长度和大行李箱的空间。

The suite after reconstruction really makes you want to stay for long. The big cloakroom, shoes area and recycle window give you enough space for hanging clothes and keeping big luggage.

细节当道，即成王道！半岛霸气！

Detail is the key for success! Peninsula Hongkong wins!

CONRAD
HOTELS & RESORTS™
康莱德酒店及度假酒店

北京康莱德酒店

30 ★★★★★

CONRAD HOTELS & RESORTS,BEIJING

Address : No.29 North Dongsanhuan Road
Chaoyang District
Beijing 100027 China
中国北京市朝阳区
东三环北路29号

Telephone : +010-6584 6000

Fax : +010-6584 6066

E-mail : Conrad_Beijing@conradhotels.com

Http : //conrad.hilton.com.cn/

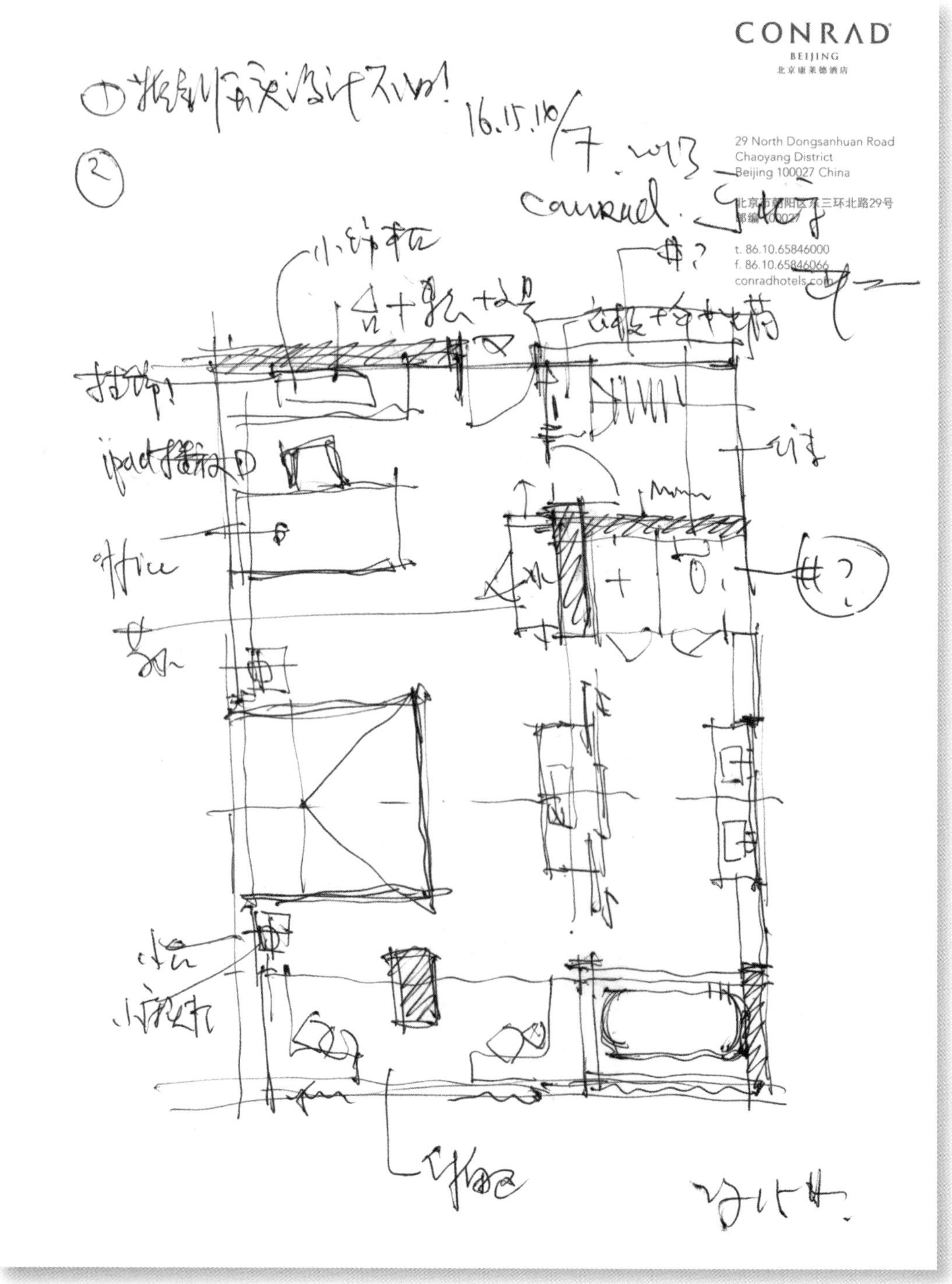

CONRAD
BEIJING
北京康莱德酒店
29 North Dongsanhuan Road
Chaoyang District
Beijing 100027 China
t. 86.10.65846000
f. 86.10.65846066
conradhotels.com

开门见床，不敢恭维。

我佩服的、熟手的、年轻才俊建筑师写出杰出的手笔，外立面形象和建筑室内内容都颇有特色，相信给北京东三环添色不少。

但，房间开门，就见到床，感觉总是怪怪的，不知当住人住会否有同感？

也许可再尝试在不影响面积效果的情况下"改变一下"，其实自认可以做得更好些，心理感觉更好，更要命的是在酒店房间门隔音差劲一般之，走廊的声音很易听到房间内的活动，不知道建筑师、室内设计师等设计团队是否有试过"小问题"作检测，再会如何决择，总之，我还始终不敢苟同，相信也会有住客提出和我一样的问题。

心理感要有时候比实际感觉更重要，这样就不会形成讨厌和抗拒，或可在书桌区前加个半透明的屏风，或改变门口入口的方式，如上海Puli酒店，

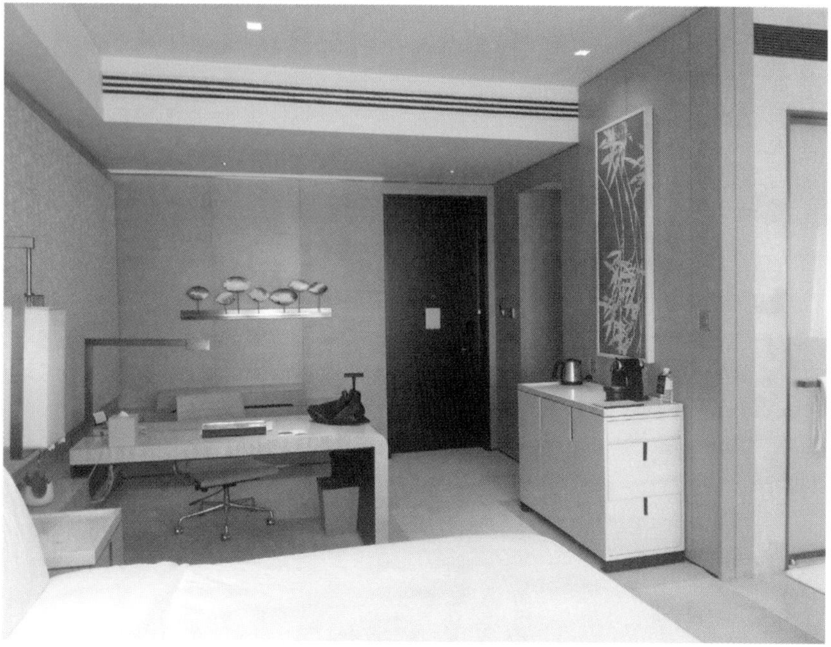

北京康莱德酒店 *Conrad Hotels & Resorts , Beijing*

Open the Door and See the Bed, personally I don't Like That.
开门见床，不敢恭维

这是出自我佩服的、炙手的，年轻才俊建筑师马岩松先生的手笔，外立面形象和建筑室内空间都颇具特色，确实给北京东三环添色不少。

I admired the masterpiece from the young and talented architect Mr. Ma Yangsong. The image of the façade and the inner space are quite individual and distinctive.It added color to the Eastern 3-rings of Beijing indeed .

但，房间开门，就见到床，感觉总是怪怪的。不知当你入住时会否有同感?

However, you can easily see the bed when opening the door, which makes you feel strange. I don't know if you will share the same feeling with me.

也动动手，尝试在不知前因后果的情况下“改变一下”。其实自认为可以做得更贴心，心理感觉更好。更要命的是现在酒店房间门隔声普遍一般般，来往的声音很影响房间内的活动。不知道建筑师、室内设计师与管理公司是否有就此“小问题 ”作权衡了。取舍和抉择，总之，我是始终不敢苟同，相信也会有住客提出我一样的问题。

Use your hand and try to "change" since you do not know the antecedent and consequence. Actually I think I can make it more intimate and feeling better. The most unsatisfactory thing is that nowadays the door isolation of the hotel is not so good. The sound from outside can easily be heard by the people inside the door. I don't know if the architect and interior designer as well as management company had discussed or considered and balanced "this small issue" or not. Take it or remove it, I have no idea. All in all, I can not agree with that and believe that some guests like me will raise the same question like me.

心理感受有时候比实际感受更重要，这样就不会形成讨厌和抗拒。或可在书写区前加个半透明的屏风，或改变门口和入口的方式，如上海PULI酒店，或许会好一些，一家之言。

The psychological feeling sometimes is more important than the real feeling. Thus you won't feel hateful or resistible, or you can add the transparent screen in front of the writing area or change the way of door gate and entrance like Shanghai PULI Hotel.Perhaps it would be better. Anyway, that is my suggestion only.

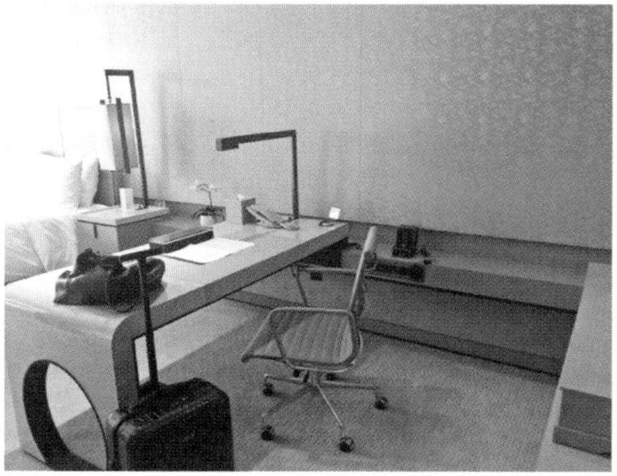

房间的价格

以前住酒店也没有太注意价格，因为这几年订房的全经由我管了，才发现到价格的差异，原来这还是很有学问的。新开的酒店异常实惠，因为在试业期间需招揽生意，有的甚至是在正式营业的一半价格左右，或是一些三、四线城市的五星级酒店，都会有相当大的地区价格差异。某些订房网站或公司还可以订到特别便宜的房间，或是夜订，或不可退换式的。天真地想想，这些同志在技巧地缩短差距之时，也为我们提供了舒适的宝石，超好！不管他了！

提前计划，少改变行程，这样就可以住到更加优惠的房间，当然，做个白老鼠住住新开的酒店也是一种试：长沙北辰洲际酒店、兰州皇冠酒店、北京康莱德酒店，刚开业时去享受啊！当然我每在一个新城市搜寻酒店都是先从开业的新旧开始，再找价格，从高到低，当然只选标准房和大床房，再筛选喜欢的品牌，如要到国外住的，则更关注网上的评论，可以说中国这几年酒店市场发展已经完全与国际接轨了，特别是价格，从一千几百到三四千元都有了，而且大家也习惯了。作为设计从业人员，当然也想从中分享到逐渐高企的价格带来我们相应的服务设计费的提升，不过我们也在做设计了，自然也知晓中各个环节都涨价了，也就不难理解房的价格的持续上升了，设计费也是"随波逐流""节节高升"了。

其他角度来看，品牌酒店的房的价格也算了一面镜子，做设计的也要如酒店样，将自己的设计作品打造成有价值的品牌，让我们的业主慕名而来，这样你的取费就会和好的一起长期，持久地上升了。

房价上升，那就沾了光了，也住不误！

House Price

房子的价格

以前住酒店也没有太注意价格，因为这几年订房的途径与方式多了，才关注起价格的差异，原来这还是很有学问的。新开的酒店异常实惠，因为在试业期间要招揽生意。有些甚至只有正式营业的一半价格左右，或选一些三、四线城市的五星级酒店，都会有相当大的地区价格差异。某些订房的网站或公司还可以订到价格特别低的房间，或先预付，或是不可退换式的。天真地想想，这些同志在有技巧地赚取差额之时，也为我们提供了简单的方便，挺好！不管他了！

I had never been so particular about the price for hotel before. I became concerned on the price discrepancy due to more selection available these years in terms of booking way and sources. I never know there is so much knowledge inside this. Usually the price for newly-open hotel will be extraordinary favorable as they need to get more business in the trial opening

period. Some price is even half of the other hotel of the same rate. Or you can select some Five-star Hotel in third-tier or fourth-tier city and they will offer quite different price. Some website specially for booking hotel or company may book very cheaper price or sometimes make prepaid or non-returned type will be cheaper. Sometimes I will be so naïve to think that these smart people can easily make money from the price gap and also in the same time they offer us the convenience. I think it is quite good! So I never care about that!

提前计划，少改变行程，这样就可以住到更加优惠的房间。当然，做做白老鼠，住住新开的酒店也是一种方式：长沙万达文华酒店，兰州皇冠假日酒店，北京康莱德酒店，刚开业就去学习学习！当然我每在一个新城市搜索酒店都是先从开业的新旧开始，之后比较价格，从高到低，当然只是标准房和大床房，再筛选喜欢的品牌。如需到国外住的，则更关注网上的评论。可以说中国这几年的酒店市场已经完全与国际接轨了，特别是价格，从一千几百到三四千的都有了，而且大家也习惯了。作为设计从业员，当然也想从中分享到逐渐高涨的价格，带来我们相应的服务设计费的提升，不然我们也白做设计了。产业链中各个环节都涨价了，也就不难理解房间价格的持续上升了。设计费也应“随波逐流”，“节节高升”了。

Make plan ahead of time and do not change your trip, which will help you get more favorable room price. Of course, you can try first like one small white mouse, which is also a very good way. For example, Wanda Vista Changsha , Crown Plaza Lanzhou,Conrad Hotels & Resorts in Beijing, I try experiencing there when it is just open! Of course, the way I search such hotels will start from the new opening one in a new city and then check the price from higher to lower, of course, just the standard room and room with bigger bed and then select the favorite brand. If you need to live abroad, you'd better concern more about the comments in the internet. We can say, hotel market during these years in China has been completely integrated with the international world, especially the price, from one thousand to three or four thousand, and people are used to it. As one designer, of course, I'd like to share the high price of the housing price which may bring us the rising of the design service charge. Otherwise, we are doing quite a lot of things for nothing. Each part in this industry chain keeps rising; we can not be difficult to understand the rising of house price. It is also the time for the design charge to be raised up "with the market condition" and keep "rising" with them.

从某种角度来看，品牌酒店的房间价格也算一面镜子。做设计的也应如酒店一样，将企业或设计师打造成有价值的品牌，让我们的业主慕名而来。这样你的取费就会与时间一起长期、持久地上升了。

From the certain prospect, the room price of the brand hotel is like one mirror. Making design is just like the way of hotel. They try to build up the company or designer into one valuable brand and attract our house owner to come. Thus the price will be kept rising with the time.

房价上升，那就轻松了，照住不误！

The house price goes up, that is easy, however, you still use it for living purpose!

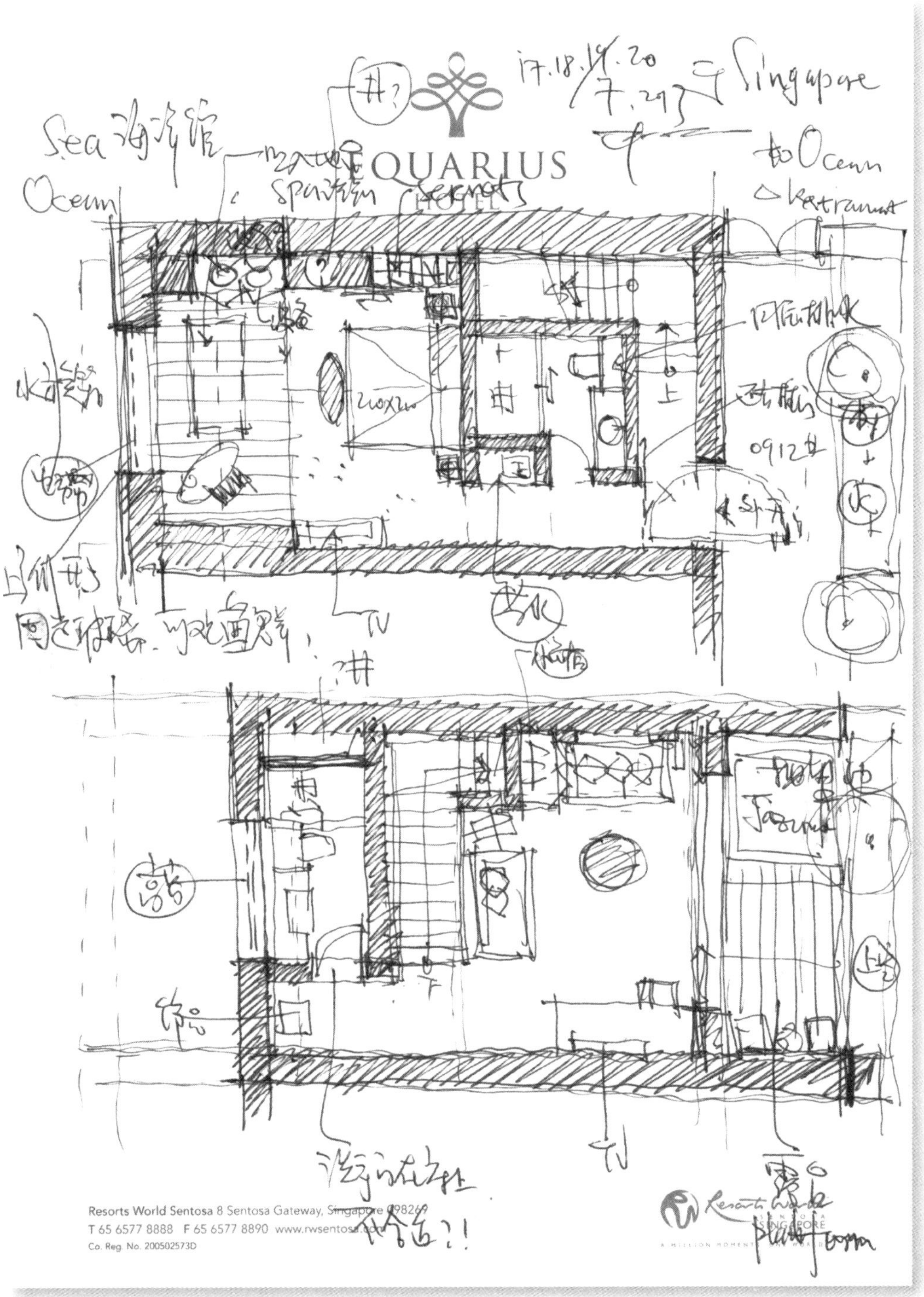
EQUARIUS
HOTEL
Sea
Ocean
Singapore
to Ocean
TV
TV
Resorts World Sentosa 8 Sentosa Gateway, Singapore 098269
T 65 6577 8888 F 65 6577 8890 www.rwsentosa.com
Co. Reg. No. 200502573D
Resorts World
SENTOSA
SINGAPORE
A MILLION MOMENTS. ONE WORLD.

Resorts World
SENTOSA
SINGAPORE

A MILLION MOMENTS. ONE WORLD.

新加坡圣淘沙名胜世界酒店

31

★★★★★

EQUARIUS HOTEL, SINGAPORE

Address : Resorts World Sentosa
8 Sentosa Gateway,
Singapore 098269
Telephone : +65-6577-8888
Fax : +65-6577-8890
Http : //www.rwsentosa.com/

新加坡的水下酒店房间

水下，并不像马尔代夫或其他海洋中真正意义的水下，而是在一个大大的海洋馆水族箱的一侧，仍然也挺震撼的。进入室内，电动帘子打开，室内灯光自然关闭（以不影响鱼儿的生活）万千鱼群游弋在眼前，蓝蓝的，浪漫得很！

这样的房间共分2层，全复式的，平面很是吸引，也不便宜（我还专门打了一下房价才五千元还不便宜的）上层是起居室，有室外平台，spa池，绿树蓝天，适合小孩居住，可以打游戏，上网，看电视，更可戏水；而下层则是主卧区，全鱼缸景观，更绝的是打开地板，可以看下面的海底深海观鱼，闭目如入海中，与鱼同游！

很有特色的一种设计，"也物尽其用"，另一侧是一间西餐厅，连通海洋馆，因为玻璃更开阔更具吸引力，上座率也颇高，午饭没地方去就浅试了一次，当然价格还是相当对得起这几片大玻璃的。

体验，不同的方式，不同的感受，或者才能激情不断的去住。

住哪？哪里都住！

The Underwater Hotel Room in Singapore

新加坡的水下酒店房间

水下，并不像马尔代夫或其他海洋中真正意义的水下，而是在一个大大的海洋馆水族箱的一侧。啊！也挺震撼的。进入室内，电动帘子打开，室内灯光自然关闭（以不影响鱼儿的生活）。万千鱼群游弋在眼前，蓝蓝的，浪漫得很！

Underwater is not like Maldives or any other offshore in the true sense of underwater but on one side of a big aquarium. Ah! Also it is quite shocking. When entering into the indoor, battery operated curtain will be opened and the light indoor will be off naturally (never affect the life of the fishes). Hundreds of thousands of fishes are swimming before you, blue and so romantic!

这样的房间共有12间，是全复式的。平面很是吸引人，也不简单（我还专门打了一下手稿才开始画正稿的）。上层是起居室，有室外平台、spa池、绿树蓝天，适合小孩居住，可以打游戏、上网、看电视，更可戏水；下层则是主卧区、鱼缸景观。更绝的是打开地板，可以有下沉的浴缸浸泡观鱼，闭目如入海中，与鱼同游！

新加坡圣淘沙名胜世界酒店 *Equarius Hotel, Singapore*

There are total 12 rooms like this and all are compound maisonettes. The plane is also quite attractive and never simple (I draw a sketch firstly before specially painting the artwork). The upper layer is the living room, the terrace, the spa pool, the blue sky with green trees, suitable for children to live. You can play games, surf the Internet, watch TV or even swim in the water. The down layer is the landscape aquarium of main bedroom,and most amazingly is that, when you open the floor, you can take a bath in the tub and watch the fishes swimming before you. Just close your eyes and swim with the fishes just like swimming in the sea!

很有特色的一种设计，也“物尽其用”；另一侧是一间西餐厅，连通海洋馆。因为玻璃更开阔，更具吸引力，上座率也颇高。午饭没地方去就将就了一次。当然价格还是相当对得起这几片大玻璃的。

One kind of design is quite distinctive which can also fully use every space it could. Another side is one western restaurant which connecting with the ocean park. Glass brings wider eyesight and more attractive with high attendance. It is lunch time and I have no choice but have to have lunch there. Of course the price is worthy of these several pieces of big glass.

体验不同的方式，不同的感觉，或者才有激情不断地去住。

Only when you have experienced in different way and feel the difference,can you have the passion to continue to do that.

住哪？哪里都住！

Where to live? Actually everywhere is worthy of living!

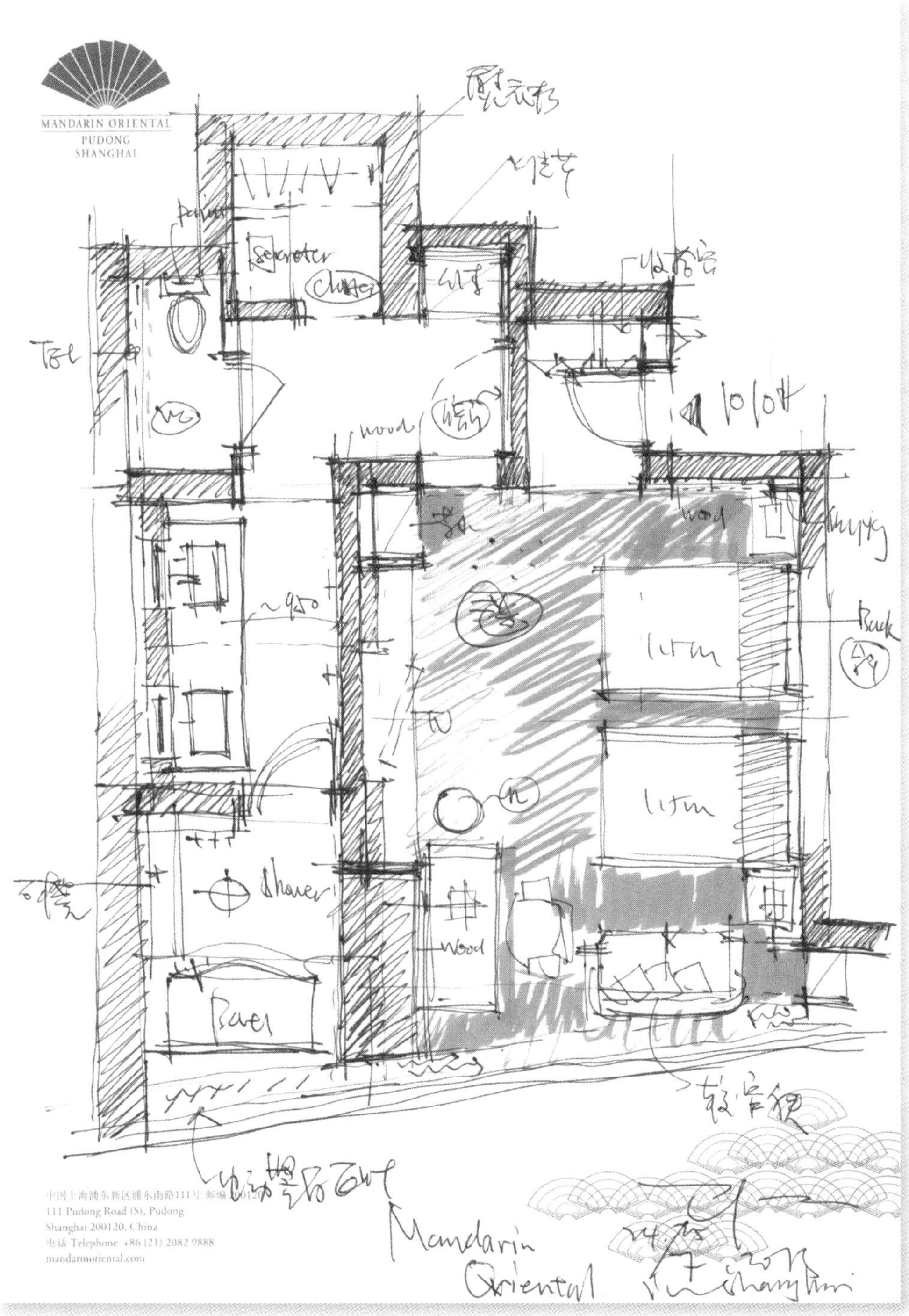
MANDARIN ORIENTAL
PUDONG
SHANGHAI
shower
wood
Mandarin
Oriental
中国上海浦东新区浦东南路111号 邮编 200120
111 Pudong Road (S), Pudong
Shanghai 200120, China
电话 Telephone +86 (21) 2082 9888
mandarinoriental.com

MANDARIN ORIENTAL
PUDONG
SHANGHAI
上海浦东文华东方酒店

32

★★★★★

MANDARIN ORIENTAL, PUDONG SHANGHAI

Address : No.111 Pudong Road (S), Pudong, Shanghai 200120, China
中国上海市浦东新区
浦东南路111号，
邮编：200120

Telephone :+86 (21) 2082 9888

Http : //www.mandarinoriental.com/shanghai/

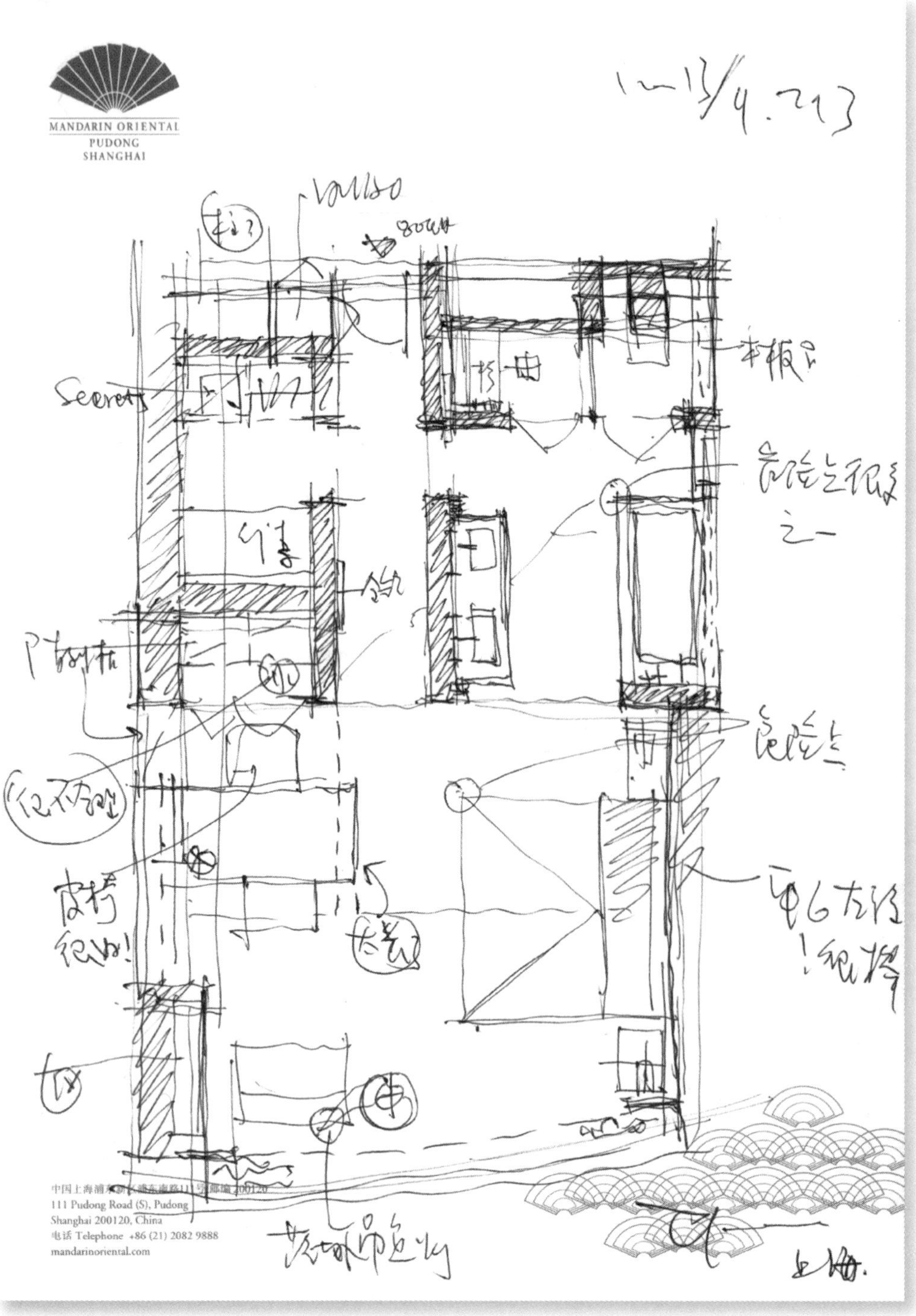
MANDARIN ORIENTAL
PUDONG
SHANGHAI
111 Pudong Road (S), Pudong
Shanghai 200120, China
电话 Telephone +86 (21) 2082 9888
mandarinoriental.com

文华·印象

听到国内同行主持设计上海的Mandarin Oriental·文华东方，十分欣喜。一则国内的专业酒店设计公司能得到认同，二则宣传后的项目会引导同业加油努力，有个可追赶的榜样。

专程去住了一下，前后两次。第一次"粗粗"看看，转转，感觉花花绿绿的，就像设计师都喜欢的色彩浓烈。可能有别于其他上海已开业或将开业的新酒店，也对；第二次感觉更深，也认识到展示了一次，体会到设计队伍的用心与水平，我去多年后的更可印证其定位的传承与独特。

这是印象，文华一向给人的沉稳、温润、高贵，这次也算是突破，希望我这种认知是对的，你认为呢？！

Impression over the Mandarin Oriental
文华印象

听到国内同行主持设计上海的Mandarin Oriental 文华东方，十分欣喜。一则国内的专业酒店设计公司得到认可，二则高品质的项目会引导同业加油向上，有学习追赶的榜样。

When I heard my peer will preside over designing the Mandarin Oriental Hotel, I am quite happy. Firstly is that now the design company can be highly recognized accepted at home. Secondly is that the high-quality project will guide our peer industry to work hard and set a very good example for our peer to follow.

专程去看了一下，前后两次。第一次“粗粗”看看、转转，感觉花花绿绿的，就像设计师初衷立意般的色彩浓烈。可能有别于其他上海已开业，或将开业的新酒店，也对。第二次感觉更深，也从头到尾参观了一次，体会到设计团队的用心与水平。或者多年后的更可印证其定位的准确与独特。

上海浦东文华东方酒店 *Mandarin Oriental, Pudong Shanghai*

I went all way to visit for two times. The first time I just overlooked it and turned around and felt that it was so colorful just like the strong color made by the designer at the first. Maybe it was quite different from other hotels which already opened in Shanghai or the new hotel that is about to open. The second time I felt more deeply. I visited from the beginning to the end and experienced the dedication and professional level of the entire design team. Maybe, many years later, we can prove its accurate positioning and uniqueness.

这是印象，文华一向给人以沉稳、低调、高贵之感。这次也算是突破。希望我这种认知是对的，你认为呢？！

This is the impression that Mandarin Oriental had shown me, calm, low-key and noble.I personally think this is a great breakthrough and I highly recognized it, what do you think?!

三亚半山半岛安纳塔拉度假酒店

33 ★★★★★

ANANTARA SANYA RESORT &SPA

Address : No.6 Xiaodonghai Road, Hedong District,Sanya, Hainan,China 572000
中国海南省三亚市
河东区小东海路6号
邮编：572000

Telephone :+86(0)898 8888 5088

Fax :+86(0)898 8888 5288

E-mail : sanya@anantara.com

Http : //sanya.anantara.com.cn/

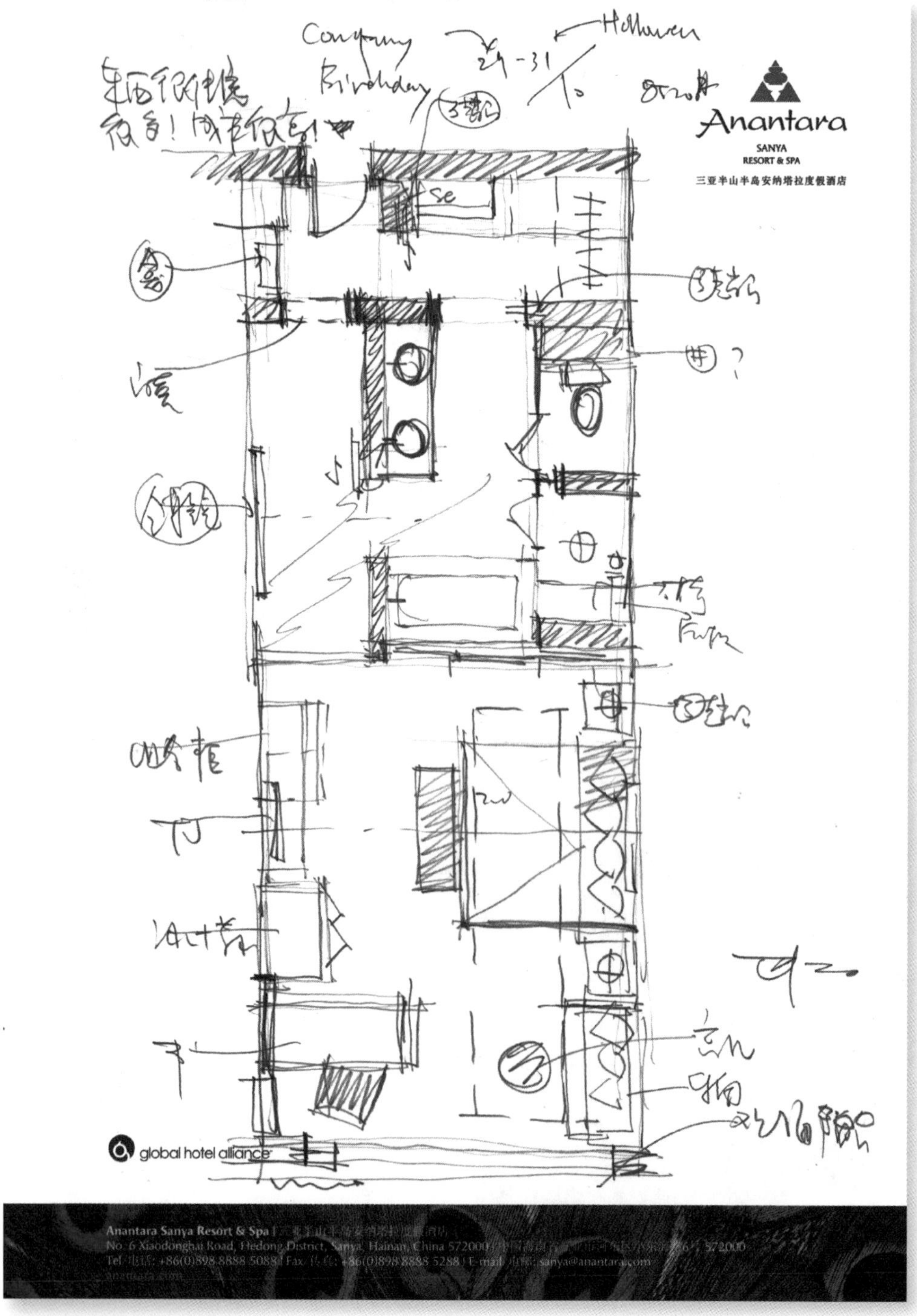

Anantara
SANYA
RESORT & SPA
三亚半山半岛安纳塔拉度假酒店
global hotel alliance
Anantara Sanya Resort & Spa

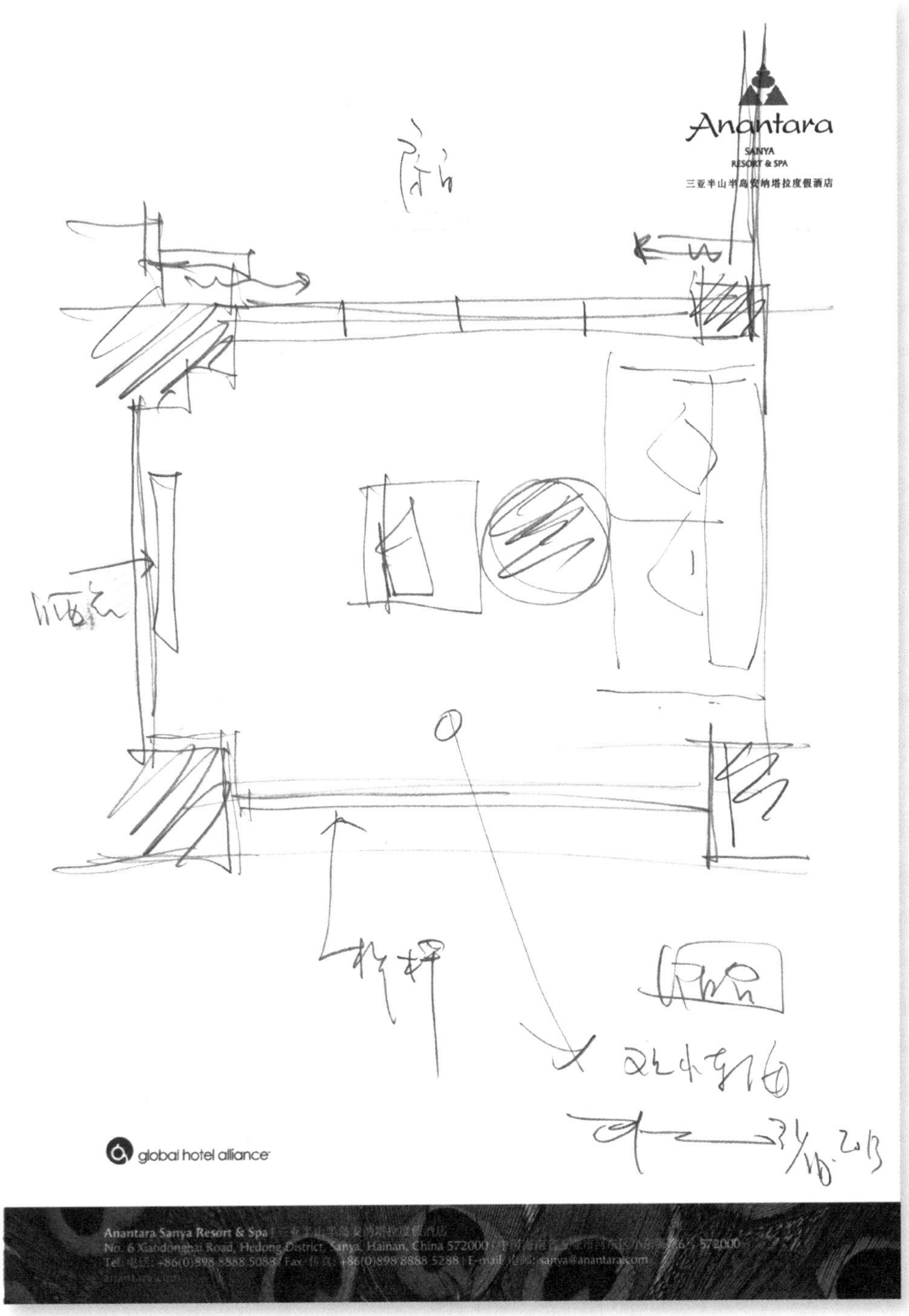
Anantara
SANYA
RESORT & SPA
三亚半山半岛安纳塔拉度假酒店
global hotel alliance
Anantara Sanya Resort & Spa
No. 6 Xiaodonghai Road, Hedong District, Sanya, Hainan, China 572000
Tel. +86(0)898 8888 5088 | Fax +86(0)898 8888 5288 | E-mail sanya@anantara.com

酒店需要什么样的风格?

每次到了巴厘，看到的以东南亚风格为主流的酒店都会在思考，其实在我们的酒店，什么样的"风格"更适合我们?!

入住的是朋友推荐住了小半月，与同一区域其他品牌的洲际酒店属同一个集团，上次《依山》曾经到了一间现代滨海风格（或称国际风）的不错的酒店，洲际，无论是主楼的客房，小组团的空间趣味，以及去小villa的错落布局及其独特的"个性"，印象不错，只是主楼过于宏大，稍嫌过闹，这间Ananatara就好多了，很舒适，入口、水景、大堂色调同样出色，建筑空间都不错，所以感觉舒服的很，最好到晚上，可能就更好了，很美!

回到房间，陈旧霉味，大片的深色令人感觉压抑了一点，减去所有的细节，包括墙面、用灯、家具都是也许只是注重细节而有特色，但没有轻松感，多了多了!

回想住过的酒店不少，"风格"之争也相当激烈，有时候会无论是业主、品牌管理者还是设计者本身都有这样的困惑，不像以前，几年前，东南亚的大气，而从仿古的做旧中式，又从东方的新中式，到近期洲际的新，瑞吉的现代豪华，悦榕庄的休闲，各能较好地找到自己的定位，但后来的"创意"很有可能往"做"差异化去研究了，或地域风格，或纯度假海滨新；又或超豪华或更偏随意"设计"式的路线；法式或中式等等。不管怎样，选择的推广，一则要结合品牌的定位和发展方向，二则要考虑建筑地点、投资额度，更要考虑市场客户和市场(经营)策略。

悦榕庄可以建在市中心，如上海、天津；安缦都可以建在山区、乡村，W酒店不也开到了海岛度假产品吗？占市场、追利润、打品牌还是主要思路，近年包括高端的安缦(Aman)、华尔道夫(Waldorf Astoria)、阿曼瑞(Amari)等奢侈小众品牌在亚洲地区也加速发展了。

"风格"嘛，根本不是酒店考虑的。如何奢侈、如何能更好地打动客人，这才是重要的。

酒店，其实真不需要什么风格的！

What Style the Hotel Need?

酒店需要什么样的风格

每每到三亚住，看到的以东南亚风格为主流的设计都会在思考，其实在我们的海岛，什么样的“风格”更适合我们呢？！

Every time when I lived in the hotel in Sanya, I keep thinking of the mainstream design when seeing the ASEAN style, actually we are in the Sea Island. What "style" will be more suitable for us?

入住的安纳塔拉位于小东海，与同一区域半山半岛的洲际酒店同属一个发展商，上次《住哪?》曾写到的一间现代海滨风格（或称国际风）的不错的酒店。洲际，无论是主楼的简练，小矮层的空间趣味，又或者小villa的错落都有其独特的“个性”，印象不错，只是与楼盘共享，稍嫌热闹。这间Ananatara就安静多了，很东南亚。入口、水景、大堂色调园林造景、建筑穿透都不错，而暗暗的感觉有点儿压抑，虽然烈日当空。可能见多了，疲劳！

Anantara Hotel where we lived was located in the Small East China Sea, which is the same area as the Semi- mountain Peninsula InterContinental Hotel. They belonged to the same developers. Last time Where To Live has ever depicted one nice hotel with modernized coastal style (or called as International Fashion). InterContinental, no matter the concise of the main building or the space interesting of the narrow storey or the unique personality of the small villa, which leaves us a very good impression. A little disappointment is that the building is co-shared and a bit noisy. Ananatara is much quieter, just like hotels in Southeast Asia. The entrance hall, water scenery, lobby-tone forest, penetration of the architecture are very nice, however, it is a bit dark and feel a bit depressed despite the strong sunlight. Maybe I have seen too much and get tired.

回到房间：陈旧、霉味，大片的深色令人感觉压抑了一点。诚然所有的细节，包括平面、用灯、造材都是大师手笔，注重细节而又经过推敲，但没有了轻松感，多了多了！

When return to the room, old and moldy, big block of deep color make people feel depressed. Actually all the details including the plane and lamp and building material are all drawn and painted by the master, detailed and particular, however, sense of relaxing was gone and too much was added.

回想住过的三亚酒店不少，“风格”之争也越发明显，相信无论是业主、品牌管理者，还是设计师都有这样的困惑。不像以前，几年前，希尔顿的大气、丽思卡尔顿的豪华中式、文华东方的新中式，到近期洲际的小清新、瑞吉的现代豪华、悦榕庄的休闲，都能较好地找到自己的定位，往后的“创意”很有可能往“微”差异化去研究了，或地域风格，或纯滨海清新，又或超豪华式，更有可能走“宫廷”式的路线，法式或中式等等。不管怎样，这样的推敲，一则要结合品牌的定位和发展方向，二则要考虑建筑地点、投资额度，更要考虑市场客户和市场（领先）策略。

三亚半山半岛安纳塔拉度假酒店 *Anantara Sanya Resort & Spa*

Looking back upon quite a lot of hotels we had lived in Sanya, "Styles" are competing more and more apparently. No matter the owner or the brand management people or designer, they all will be puzzled and at a loss. Not like before, several years ago, the magnificence of Hilton, the luxurious Chinese style of Ritz-Carlton, the new Chinese style of Mandarin Oriental, small refreshing Intercontinental recently, modern and luxurious Raj, leisure Banyan Tree , all you can find their own positioning. The "idea" later is likely to turn to "micro" differentiation or the regional style feature or pure coastal refreshing style or super luxurious type, more likely to go the "imperial" way, French or Chinese, etc. Anyway, through such thinking, one point is that you need to confirm the developing direction by combining all the brand positioning. Secondly, you need to consider the construction site, investment amount. Moreover you need to consider the market and customer (priority) strategy.

悦榕庄可以建在市中心，如上海、天津。希尔顿可以建在山区、树林。W酒店不也开了不少海岛度假产品吗？占市场，追利润，拓品牌还是主要思路。近年包括顶端的安缦（Aman）、华尔道夫（Waldorf Astoria）、阿曼尼（Armani）等奢侈小众品牌在亚太地区也加速发展了。

Banyan Tree can be built in the city downtown, like Shanghai, Tianjin. Hilton can be built on the mountain or in the wood. Didn't W Hotel also open so many vacation products for the island? Occupy the market and pursue the profit. The point for promoting the brand mainly lied on the tactic and idea, recently including the top hotel like Aman, Waldorf Astoria, Armani and some luxurious small brand are speeding up to promote their business in Asia-Pacific region.

"风格"嘛，根本不是主要考虑的。如何奢华，如何能更准确地打动客人，这才是重要的。

As for the "Style", it is absolutely not the main factor to consider. How to make it more extravagant and luxurious and how to make it more accurate to impress the customers, which is the point.

酒店，其实真不需要什么风格的。

As for the style of the hotel, actually no one knows for sure which style is the best.

三亚半山半岛安纳塔拉度假酒店 *Anantara Sanya Resort & Spa*

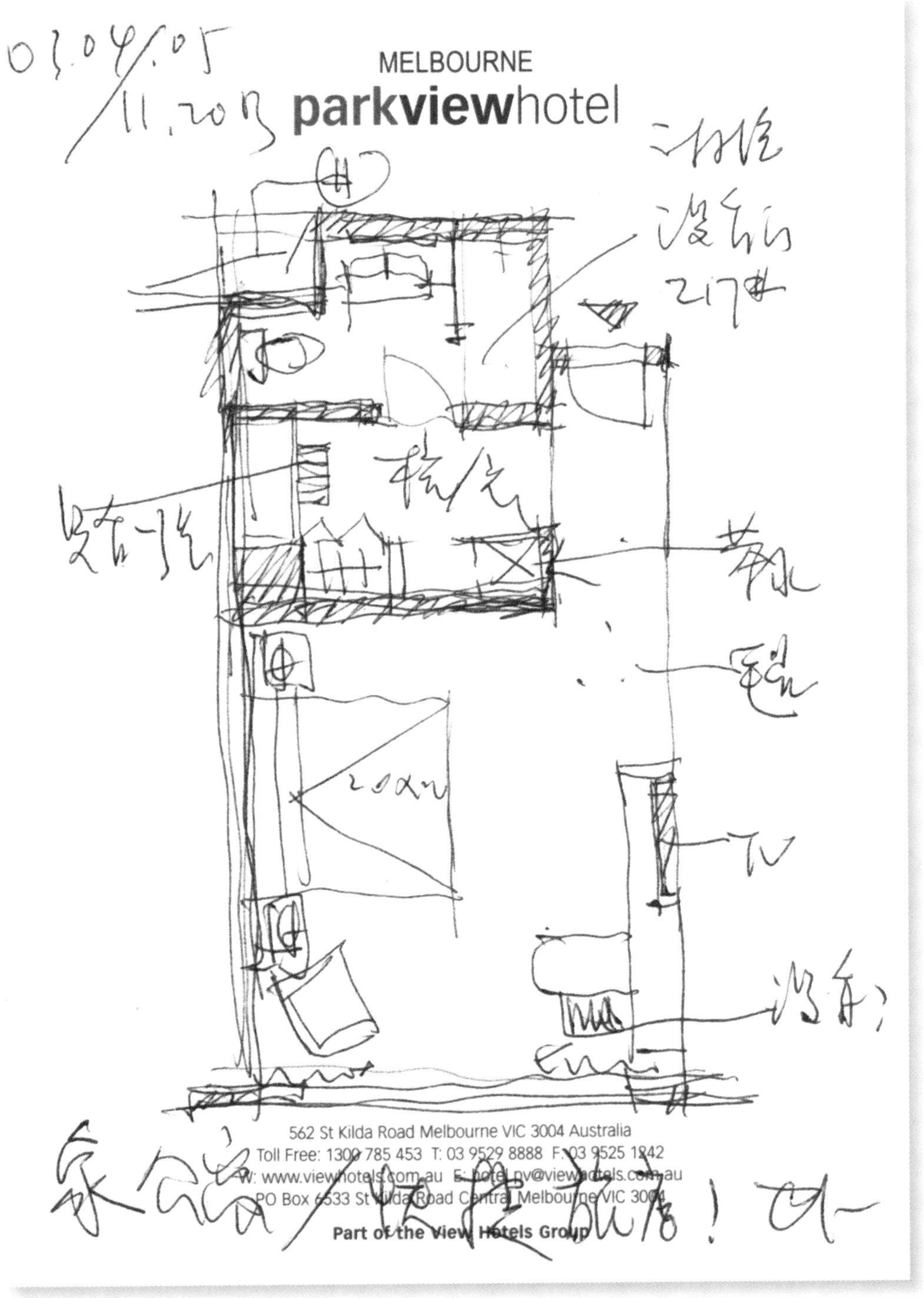
MELBOURNE
parkviewhotel
562 St Kilda Road Melbourne VIC 3004 Australia
Part of the View Hotels Group

viewhotels
AUSTRALIA

墨尔本美仑大饭店

34

MELBOURNE PACKVIEW HOTEL

Address : No.562 St Kilda Road Melbourne, Victoria 3004 Australia
Telephone :+61 3 9529 8888
Fax :+61 3 9525 1242
E-mail : hotel.pv@viewhotels.com.au
Http : //www.viewhotels.com.au/

Holiday Inn

凯恩斯假日酒店

35 ★★★★★

HOLIDAY INN CAIRNS

Address :No.121-123 The Esplanade & Florence ST PO Box 121 Cairns 4870

Telephone : +61-7-40506070

Fax : + 61-7-40313770

Http : //www.holidayinn.com/

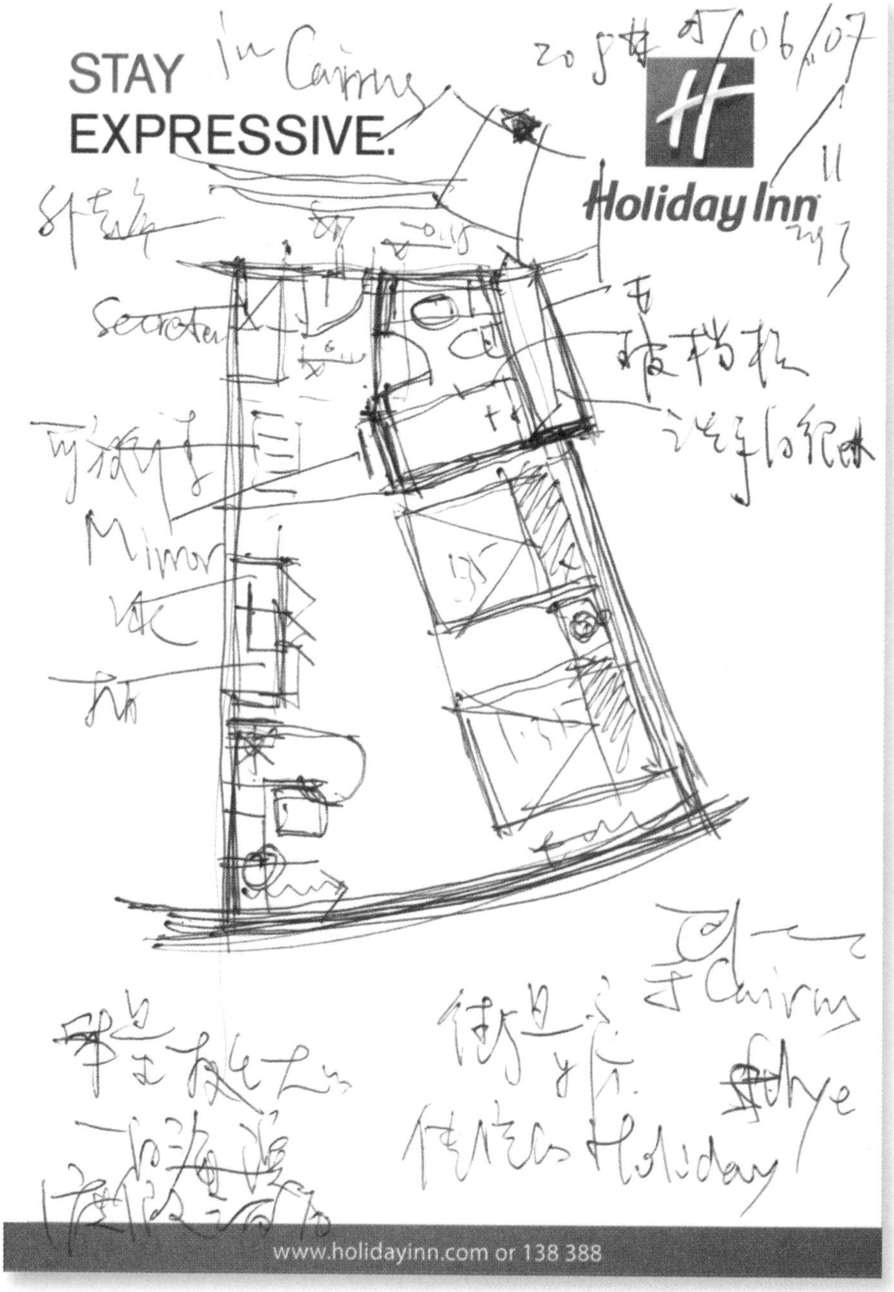
STAY
EXPRESSIVE.
Holiday Inn
www.holidayinn.com or 138 388

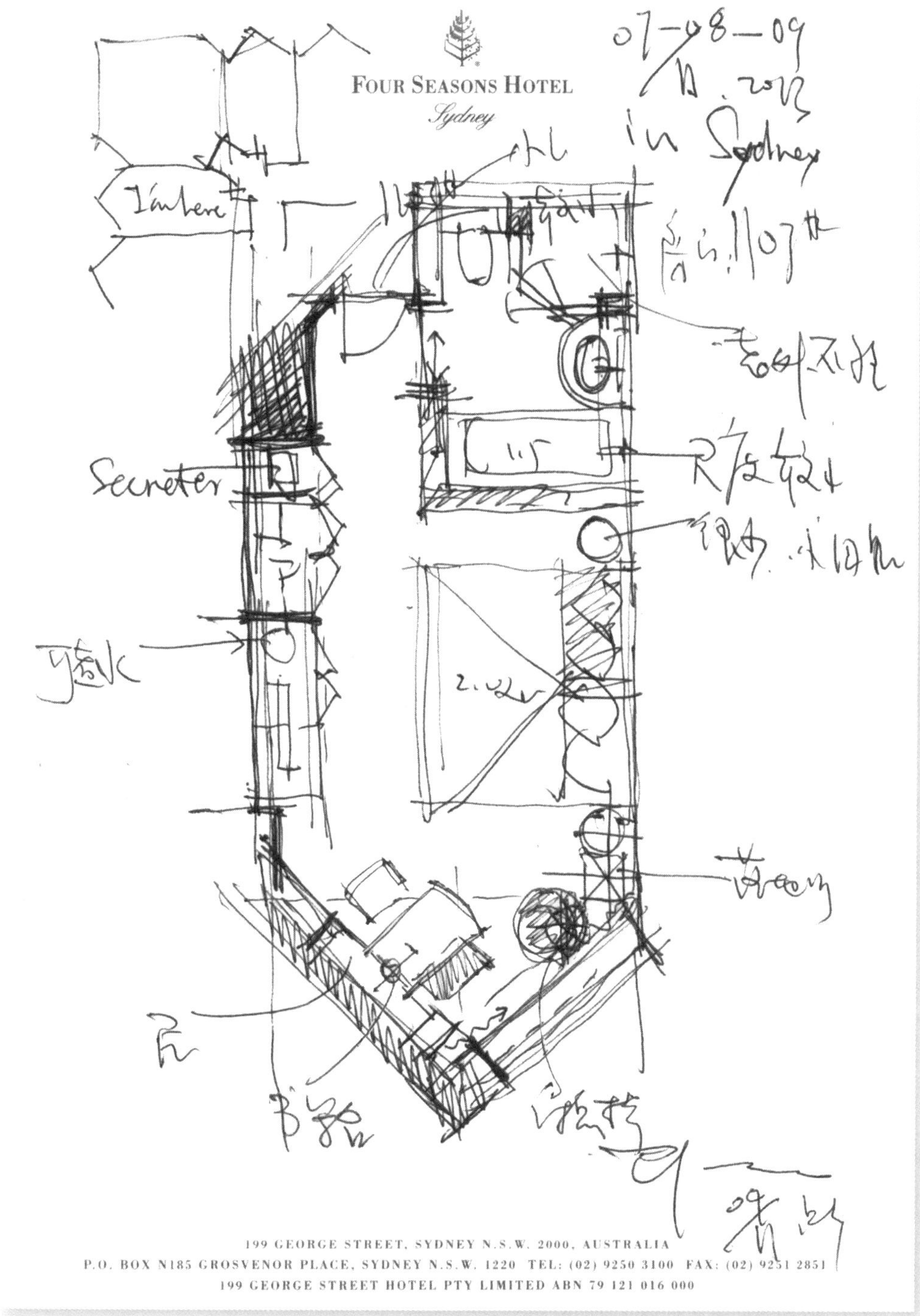
FOUR SEASONS HOTEL
Sydney
07-08-09
in Sydney
Secreter
199 GEORGE STREET, SYDNEY N.S.W. 2000, AUSTRALIA
P.O. BOX N185 GROSVENOR PLACE, SYDNEY N.S.W. 1220 TEL: (02) 9250 3100 FAX: (02) 9251 2851
199 GEORGE STREET HOTEL PTY LIMITED ABN 79 121 016 000

FOUR SEASONS HOTEL
Sydney
悉尼四季酒店
★★★★★

36

FOUR SEASONS HOTEL, SYDNEY

Address : No.199 George Street
Sydney NSW 2000
Telephone :+61 (2) 9250-3100
Fax :+61 (2) 9251-2851
E-mail : hotel.pv@viewhotels.com.au
Http : //www.fourseasons.com/sydney/

灯光

悉尼是这次公司中层澳洲游学的最后一站，选了当地最好的其中一间酒店：四季酒店，步行几分钟就可到了海岸，沿着弯弯曲曲的港湾看到侧面。美如斯的悉尼歌剧院，蔚蓝的海水倒影着其如女子般的样子！真美！

酒店位于黄金地段的佐治街，可能这个原因吧，酒店室内并不是想象中的宏伟，但一切都是精心设计的，大堂的色调、现代家具、经典的艺术品，特色灯饰的镜子，客房的艺术细部，然而最特别的是它的灯光设计。

和亚洲酒店相尽奢华材料的堆砌不同，这间四季酒店更加平实、亲民。公共区域大面积的木饰面，结合柔和、随意而泛黄的灯光，温暖而有层次；到了客房则更是体现运用的巧妙，可能是建筑的先天制约，客房较窄狭，层高更是特别、特别的矮，连窗帘杆也收到[illegible]去了，灯光在这里起到关键、神奇的作用，没有的不够，而绝不多余滥用：书柜的照明漂亮，而床头的舒适，安坐阅读中，多样要求的灯光都满足，结合大暖色的搭配，简直是完美设计之作，回国后还找到以完美的例子向同事们讲解其灯光运用之精妙。

不枉我们在这里呆了三天，感受到不同时段的光和灯光之变化，也算是这趟澳洲游完美之一站!

Light
灯光

悉尼是这次公司中层澳洲游的最后一站，选了当地最好的其中一间酒店：四季酒店，步行几分钟就可到海旁。隔着百来米的港湾看到侧面美如斯的悉尼歌剧院，深蓝的海水倒映着其如处子般的样子！柔美！

Sydney is the last leg of this Australian trip for the middle level of staff in our company. Choose the best one in the local area, Four Season Hotel, which is just several minutes away from the sea. From the harbor in hundreds of meters away and you still can see the side of the beautiful Sydney Opera House. The shadow reflecting upside down on the deep blue sea is very like the appearance of a virgin! Soft and sweet!

悉尼四季酒店 *Four Seasons Hotel , Sydney*

酒店位于黄金地段的佐治街，可能这个原因吧，酒店室内并不是想象中的宏伟，但一切都是精心设计的。大堂的色调，形式简单，经典的采光井，特色惊喜的餐厅，多层的共享环廊，然而印象特深的是它的灯光设计。

The hotel is located at the golden section of George Street, this may be the reason, the interior of the hotel is not so magnificent as you can imagine before. However, all are specially designed. The tone of the lobby, simple form with classic lighting well, amazing canteen, more storeys of co-shared corridor, however, the most deepest impression on my mind is still its lighting design.

和亚洲酒店极尽奢华、材料的堆砌不同，这间四季酒店可谓平实、亲民。公共区域大面积的木饰面，结合看似随意而泛黄的灯光，温暖而有层次。到达房间更是体现运用的巧妙，可能是建筑的先天制约，房间较窄狭，层高更是特别特别的矮，连空调机也收到洗手间去了。灯光在这里起到关键神奇的作用，该有的不少，而绝不多余滥用：茶水柜的照明够亮，而休闲区够柔，写字区够集中，多样要求的洗手间够混搭，结合大暖色的搭配，简直是完美而简约之作。回国后还专门以它为例向同事们讲解其灯光运用的精妙。

Different from the extreme luxury and material stacking of Asian Hotel, the Four Seasons Hotel can be called plain and easygoing. Public area is equipped with wooden decoration surface, combined with the yellow light, warm and layering, and more skills are applied to make the light fully reach the room. Maybe it is limited by the inborn building, the room is quite narrow, and the higher is especially short. Even the air-conditioner can be stored into the washing room. Light plays an important role and miraculous function here. No less but never too much: the illuminant of the tea cabinet is bright enough and soft in the leisure zone; the writing area is concentrated enough; washing rooms with various requirements are mixed together. Combining with the warm color, it is simply the coordination between the perfect and simplicity. When came back home, I also specially took it for example and explained the wonderful and perfect application of the light to my colleagues.

悉尼四季酒店 *Four Seasons Hotel , Sydney*

不枉我们在这里待了三天，感受到不同时段的阳光和灯光的交融，也算是这趟澳洲游完美的一站！

It was worthy of us to stay here for 3 days and felt the fusion of the sunlight and light in different time section, which could be called the most perfect leg of the Australian Trip.

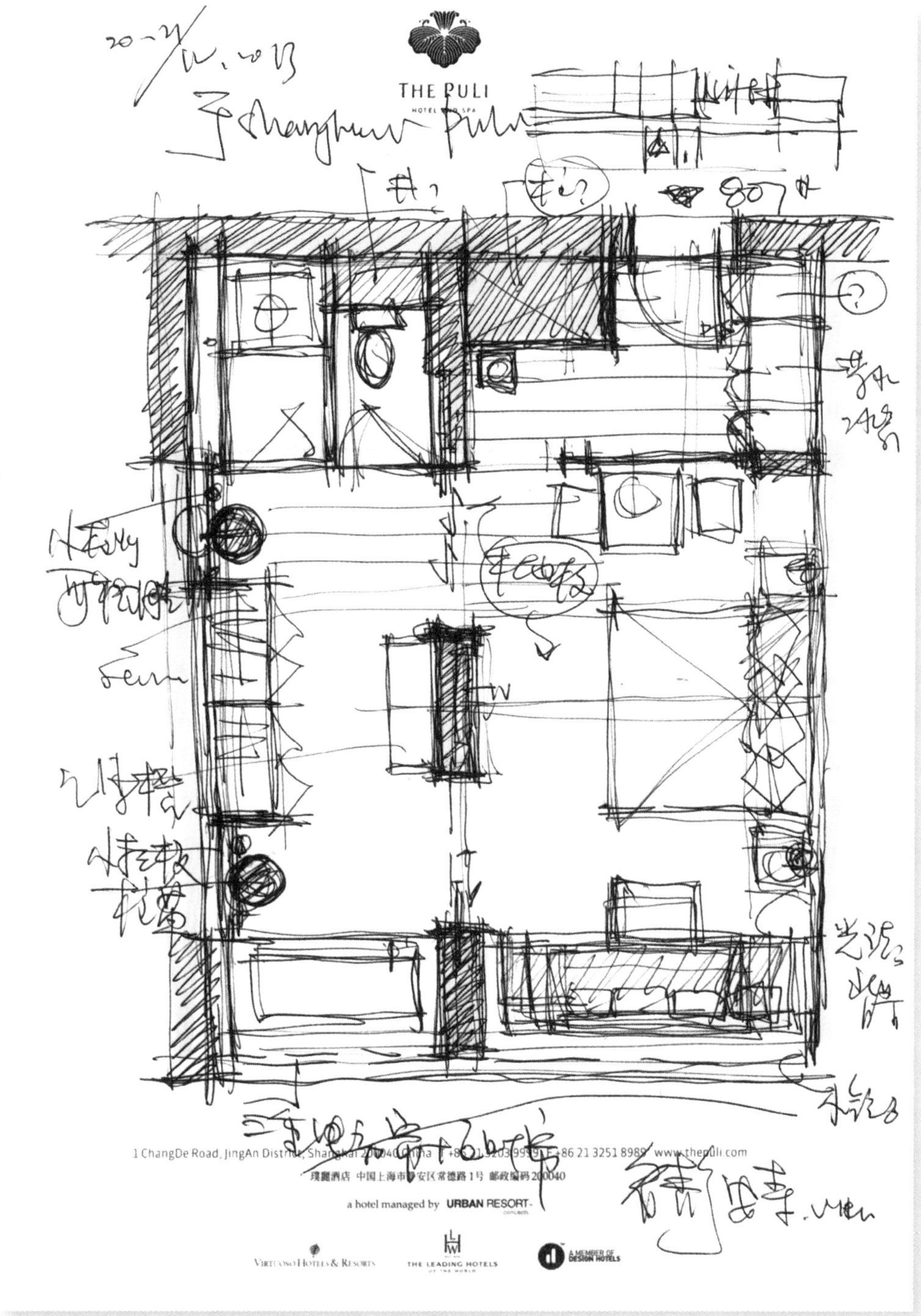
THE PULI
a hotel managed by URBAN RESORT concepts
VIRTUOSO HOTELS & RESORTS
THE LEADING HOTELS OF THE WORLD
A MEMBER OF DESIGN HOTELS

THE PULI
HOTEL AND SPA
上海璞丽酒店
★★★★★

37

THE PULI HOTEL & SPA, SHANGHAI

Address : No.1 ChangDe Road,
JingAn District
Shanghai 200040 China
中国上海市静安区
常德路 1 号
Telephone : +86 21 3203 9999
Fax : +86 21 3251 8999
Http : //www.thepuli.com/zh/

21/12.2013.

THE PULI
HOTEL AND SPA

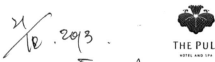

璞丽. From Shanghai To Canton.

关于Puli的回顾:

刚刚我也不太记得Puli的经过，从设计到建造的全过程，好像也就那时候认真地一些东西、空间、光的气氛也带进去了。中间有设计和施工的过程，客房和餐厅、spa、酒吧，这些都要想了一下，细节的考虑都在不同的地方成为(酒店)的独特之处。

平面很简单很简单，没有什么特别之处，是东西向进入的一条走廊的设计也很简单了的，但细节下来才感受到它难做的重要和细腻之处。从这些的酒店里面中能感受到建筑之外，也感受很多Fans多年相处，再到对方的设计感情之间，这样……共同的设计的投资，不是一种……不是一种材料

1 ChangDe Road, JingAn District, Shanghai 200040 China T +86 21 3203 9999 F +86 21 3251 8989 www.thepuli.com
璞麗酒店 中国上海市静安区常德路1号 邮政编码200040

更是令人向往，其高服务也是一流的一大特色设备，如果以细节而言，不论是精致的设计与陈设，体贴之至的服务，令客人有宾至如归的感受，全方位的空间，这次下来的住宿心情可真是愉快，所以我觉得也是极好的。

好的国家的酒店，或是说以我们的角度去评价一下好的酒店，也是一种享受生活和品味的艺术的一种体验，也是我们对生活的好的享受！

21

About Public Praise of Puli Hotel

关于Puli的口碑

刚开业不久就去Puli看过，从头到尾参观过。后来也去那里消费了一些空间，如餐厅、酒吧等等。也带过同事、甲方去交流和观赏过，包括房间、SPA、泳池区。这次慢慢地住一下，细细感受其核心区域房间的过人之处。

I have been to Puli Hotel when it was just opened. I walked around the whole hotel, and used to consume in some areas such as canteen and bar. I also went there with my colleagues or clients, visiting hotel rooms, SPA and swimming pool. This time I lived here and felt all the advantages in the core part of hotel, the room.

平面很简单，没有什么特别之处。表面看，包括入口的半透屏风的设量也颇受争议，但细细画下来才感受到其难得的严谨和细腻，是住过的酒店里少有的极致之作，也难怪很多Fans多年拥戴。再到灯光的设计与控制运用，选材、造型的简之而极简，不多一条线，不多一种材料，更是令人佩服。当然也是高投入。一间房间一大特色设备，灯具和饰品无一不是精品设计与订制。你想想这可是五六年前的成果啊，而且选用的是全木地板。这几年下来的维护心情可想而知，现在去看来也是挺好的。

The layout is very very simple, without anything special. For the first sight, the entrance, including the half transparent screen, is controversial. But until I finished drawing the layout carefully, I finally found the strict and exquisite details. It is really a perfect work among so many hotels I have ever lived. It's no wonder there are so many fans of Puil for years. Back to the light design and using, the material and decoration are very simple, without any extra line or extra material. It's so admirable. Of course another special feature of this room is high cost. All facilities, lights and arts are creative custom items. You must know they were all made five to six years ago, as well as the wooden floor for all the room. But under the good protection, now everything is still all right.

上海璞丽酒店 *The Puli Hotel & Spa, Shanghai*

好好感受的酒店，或者过了几年再去住一下好的酒店，也是考验其管理水平的一种好的体验，也是我的《住哪?》好的题材!

Feeling a hotel carefully, or going back to a good hotel after a few years, you can judge the management of a hotel. It's also a good subject for my book Where to Live!

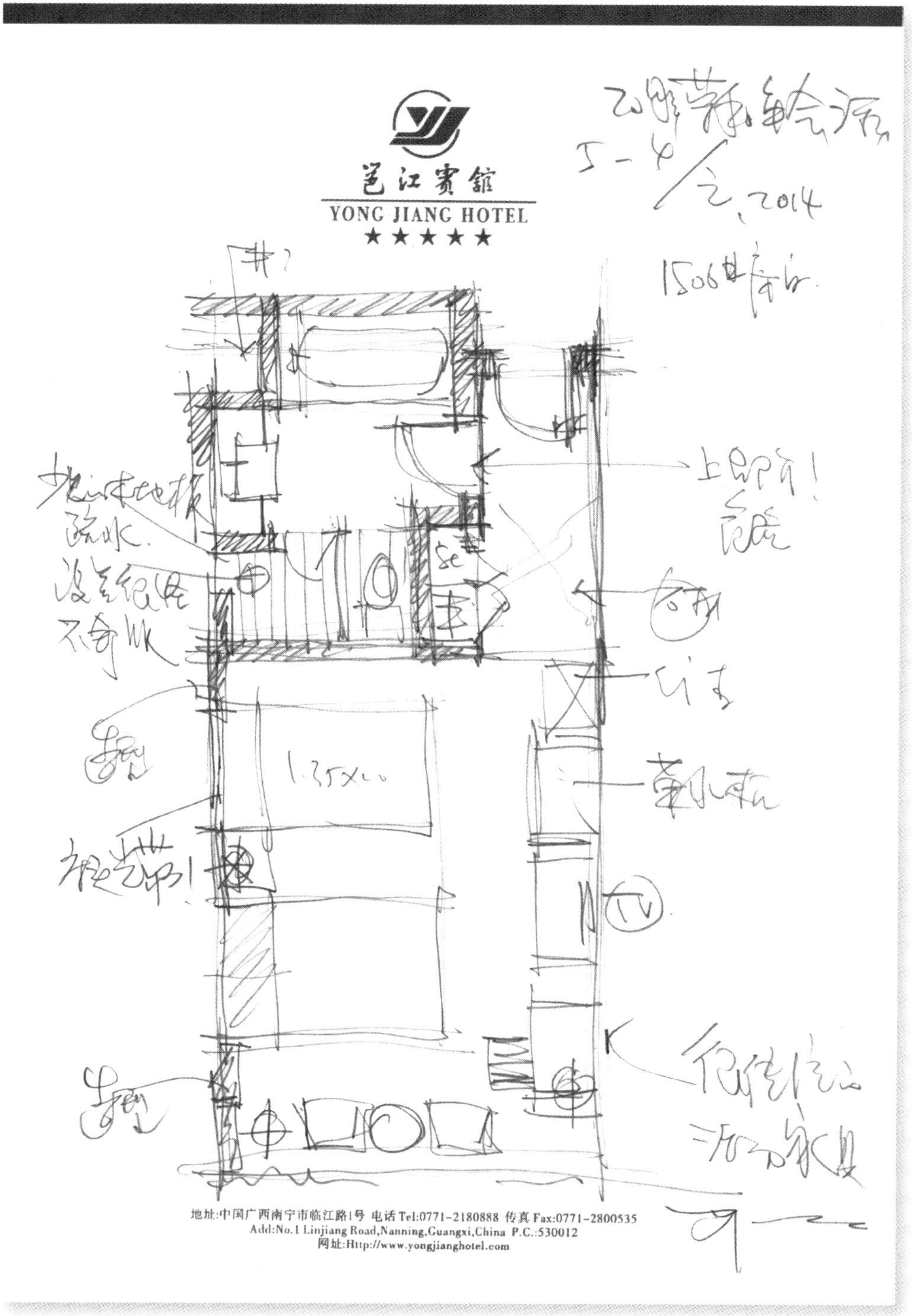
邕江賓館
YONG JIANG HOTEL
地址:中国广西南宁市临江路1号 电话 Tel:0771-2180888 传真 Fax:0771-2800535
Add:No.1 Linjiang Road,Nanning,Guangxi,China P.C.:530012
网址:Http://www.yongjianghotel.com

邕江賓館
YONGJIANG HOTEL
南宁邕江宾馆

38

YONGJIANG HOTEL, NANNING

Address : No.1 Linjiang Road,Nanning, Guangxi,China P.C.:530012
中国广西南宁市临江路1号
Telephone :+0771-2180888
Fax :+0771-2800535
Http : //www.yongjianghotel.com/

有特色的邕城五星级酒店

第三次入住了南宁的五星级酒店，除了前住的万豪品牌外，其他两间均是当地品牌，包括这次入住的邕江宾馆，是我们的客户荣和集团属下的企业。

该酒店占尽了地理优势和历史优势，有其独特的人文吸引力，因为是客户的接待礼仪上，服务上都是超乎想象的好，之前从微信上也获悉部分客房已重新翻修，入住后感受到睡眠区的新面貌，灯光、墙面、家具等，浴室区域依旧，就像入住到新开的香港半岛酒店一样，翻修了睡区、更甚者，整个浴室都是抬高了一级的，淋浴区，坐便区都用了木地板，为了预防漏水，但格局依旧可能受建筑的空间限制了吧。

甚至如果问我们可以怎么做呢？酒店改造的学问还是不少的呢。少有客户如广州的白天鹅酒店全部停业三年翻新，不惜出资的大手笔；也有像我们合作多年的东莞宾馆酒店客房层全停业，从设计到重新营业，七个月完成的决心，所以我们的客户！

鉴江宾馆因此改造而未能与以往媲美，也算遗憾。从地域性的酒店改造来看，合理投入、巧妙定位，分步走是必要而可行的，也能够有效地提升地区性酒店的形象。当然作为设计师更应了解业主的意向和市场的要求也至关重要，不能什么作品都出，也要有意思！

“特色主题”连锁民企（非国际品牌酒店管理公司管理的）主导级酒店的不二选择，不然会很快被外来的“狼”硬生生地吞掉的！

Characteristic Regional Five-star Hotel
有特色的区域五星级酒店

第三次入住广西南宁的五星级酒店。除之前住的万豪品牌外，其他两间均是当地品牌，包括这次入住的邕江宾馆，是我们的客户荣和集团属下的企业。

This is my third time living in this Five-star Hotel in Nanning, Guangxi. Except for Marriott brand, other hotels are all local brand including the Yongjiang Hotel where I lived for this time, which is one enterprise subordinated to our customer Ronghe Group.

老酒店有它的地理优势和历史优势，有其独特的人文吸引力。因为是客户的接待，礼仪上、服务上都是超乎想象的好。之前从微信上也获悉部分的客房已重新翻修，入住后感受到睡眠区的新形象，灯光、墙面、家具等等。洗手间区域依旧，就像入住翻新后的香港半岛酒店一样。单单留下了湿区，更甚者，整个洗手间是抬高了一级的，淋浴区、坐便区都用了木地板，为了方便流水，很脏很旧，可能是受建筑的局限制约吧。

Old hotel had its geographic advantage and history advantage as well as its unique cultural attraction to impress its customers. As we are received by our customer and the service is extraordinary better than we can imagine in terms of etiquette or service. I can see some of the guestrooms from Wei Chat and know that this hotel had been renovated. When I lived here, I could feel the new image from the light, wall, furniture, etc. Washing room area is still the same as the old days, just like the newly renovated The Peninsula Hongkong. Only the wet area left, even the entire washing room is upgraded for one level. The shower area and toilet are paved with wooden floor for the convenience of water flowing. It is dirty and very old. Maybe it is due to the limitation of the architecture.

想想如果交给我们可以怎么做呢？酒店改造的学问还是不少的呢。少有客户如广州的白天鹅酒店全部停业三年翻新，不做生意的大手笔，也少有像我们全程参与的东莞宏远酒店客房层全停业，从设计到重新营业，七个月完成的决心。谢谢我们的客户！

Just thinking that, If this decoration job is transferred on us, what can we do? The renovation job is quite particular and so many to learn. Actually very few hotel will do like White Swan Hotel Guangzhou before. They would rather stop business 3 years for new renovation, also seldom hotel like our customer Dongguan Hongyuan Hotel that they could stop business for all the guestrooms. Start from design to re-open, they made great resolution to finish it within 7 months. I also here to express my thanks to our customer!

邕江宾馆经过改造而未能与时并进，也算遗憾。从地方性的酒店改造来看。合理投入、巧妙定位、分步走是必要和可行的，也能慢慢地提升地区性酒店的形象。当然作为设计师更应了解业主的意向和市场的要求也至关重要，不然的你付出也是自费的！

Although Yongjiang Hotel also have been renovated, however, what a pity that it still can not keep abreast with the time. From the transformation of the local hotel, it is necessary and feasible to invest reasonably and position accurately and then go forward step by step, which can also enhance the profile of the hotel in the local area slowly. Of course, it is always vital and important that the designer should better understand the intention of the owner and the market demand. Otherwise all the pay and hardworking will be just in vain.

“特色”应当是这类民企（非国际品牌酒店管理公司管理的）五星级酒店的不二选择，不然会很快被外来的“狼”硬生生地吞掉！

"Characteristic" should be the only choice for these private-own Five-star Hotels (not managed by International Branded Hotel & Management Company). Otherwise they will be swallowed by the wolves of his competitive counterparts from outside!

南宁邕江宾馆 *Yongjiang Hotel , Nanning*

长隆横琴湾酒店

39

CHIMELONG HENGQIN BAY HOTEL

Address : Fuxiang Bay, Hengqin New Distrcit, Zhuhai City, Guangdong 519031, People's Republic of China
中国广东省珠海市横琴新区富祥湾

Telephone : +86 756 299 8888

Fax : +86 756 299 8800

Http : //www.chimelonghengqinbayhotel.com/

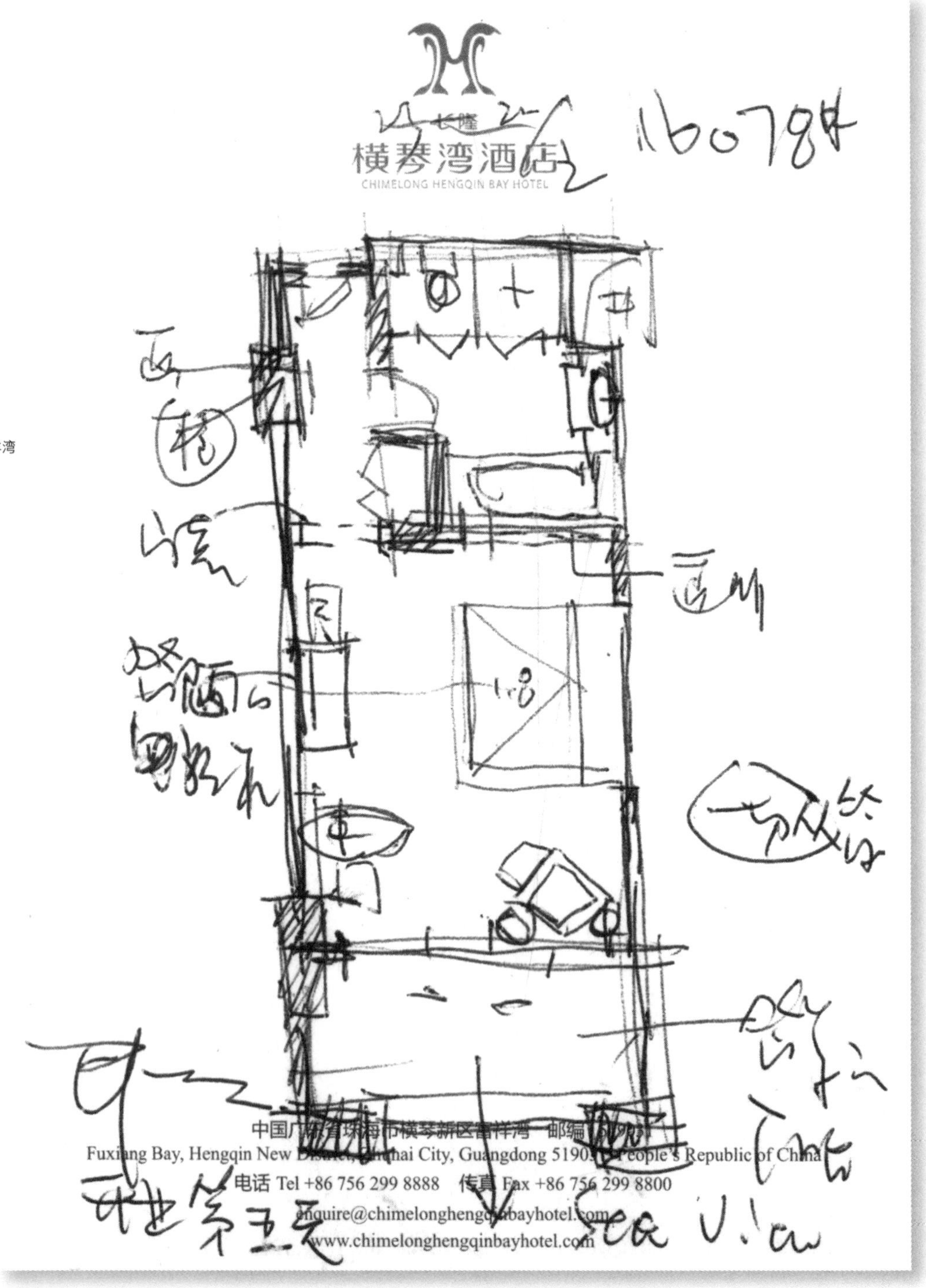
横琴湾酒店
CHIMELONG HENGQIN BAY HOTEL
中国广东省珠海市横琴新区富祥湾
Fuxiang Bay, Hengqin New District, Zhuhai City, Guangdong 519031 People's Republic of China
电话 Tel +86 756 299 8888 传真 Fax +86 756 299 8800
enquire@chimelonghengqinbayhotel.com
www.chimelonghengqinbayhotel.com
Sea View

主题公园里的酒店

开业没几天，到了迪拜住棕榈岛上的亚特兰蒂斯酒店看看。住久，周边还是一片大工地，应当说整个棕榈岛都是一个大工地。酒店在端头，说的是在尽头，相信日后的环境还是不错的。用他们的话说：未来是全世界最高级、游乐设施最丰富也最富想象力的海洋动物主题公园。

酒店开得仓促，只有部分开业，准确说只有很少的一部分投入使用。仿埃及城堡的酒店大堂还是挺壮观的，两边两侧的观水“海狮”，大海豚的环形玻璃，万鱼汇聚的顶棚，抬头望去，宛如置身浪漫旖旎的热带海底。

到了房间就体验到了埃及三星级的定位，全开业后客房多达1888间的客房长廊一望的摆设数量可见，接待能力强。一个多月后再去光顾的时候，已是人满为患了。

主题公园里的酒店去的不多，不知道像迪士尼的那套酒店是不是也像这种样式。但是就如埃及陶瓷古城的那种方式印象深一些，有不同的文起区域：海洋馆、环球马戏城、赌场、商业街，而配套的酒店更是高中低档都有，有四星的Hard rock，更有以spa为主题的酒店，不带任何购买的商业与赌场、度假区；也包括能入住的"海底"（例如上文在海洋馆旁大的水族箱侧）房间酒店叫"EQUARIUS"。各个区域以环回电动穿梭车连接，这样的规划就能满足不同需求的客人，相信它在建成后的长隆会有更多的惊喜。

我们期待着。

Theme Park Hotel
主题公园里的酒店

开业没几天，去了珠海横琴岛上的长隆横琴湾酒店看看、住住。周边还是一片大工地，应当说整个横琴岛都是一个大工地。酒店在端头，澳门近在咫尺，相信日后环境还是不错的。用他们的话说："未来是全世界顶级、游乐设施最丰富也最有想象力的海洋动物主题公园。"

Opening just a few days, I went to the Changlong Hengqin Bay Hotel to have a look, which is located at Zhuhai Hengqin Island.The surrounding area is quite a big site. It should be said the Hengqin Island is a large site. The hotel is at one end that quite close to Macao. I believed that the environment's future should be very good. They said that it would have the most promising future in the world. Also it would have the richest recreational facility and the most abundant and imaginative Ocean Creature Theme Park.

长隆横琴湾酒店 *Chimelong Hengqin Bay Hotel*

酒店开得仓促，只有部分开业，准确说只有少少的一部分投入使用。公共区域及酒店大堂还是挺童话的，雨篷两侧的戏水“海狮”、大海豚的环形立柱、万鱼汇聚的顶棚，抬头望去，宛如置身浪漫缤纷的热带海底。

The hotel was opened in haste and only opened partially. Accurately speaking, only a small part was put into use. Public area and the hotel lobby is quite like fairy tale, the paddling "Sea Lion" at both sides of the awning and annular column of big dolphin, ceiling of hundreds of thousands of fishes converged. When we looked it up, it was just like living in the romantic colorful tropical sea.

到了房间就简陋多了，像三星级的定位，全开业后有多达1888间的客房。长隆的一贯的手法是数量为先，接待能力强。一个多月后再去光顾自助餐，已是人满为患了。

When arriving at the hotel, you will find the room is much humble, like such kind of three-star hotel. The whole city has more than 1888 guestrooms, and Changlong have been taking the lead in large quantity of guestrooms and strong in reception capacity. One month later, when I visited the buffet, it was already so crowded at that time.

主题公园里的酒店去得不多，不知道像迪士尼的配套酒店是不是也像这种方式。倒是新加坡圣陶沙区域的规划方式印象深一些，有不同的主题区域：海洋馆、环球影院、赌场、商业街，而配套的酒店更是高、中、低档都有，有四星的Hard Rock，更有以Spa为主题的酒店，也包括我们入住的“海底”（实际上是在海洋馆最大的水族箱位置的酒店：EQUARIUS HOTEL），不带任何嘈杂的商业与赌场，纯度假。各个区域以环回电动穿梭车连接，这样的规划就能满足不同需求的客户了。相信完整建成后的长隆会有更多的惊喜。

I have seldom been to the hotel in the theme park. I didn't know if the hotels inside Disney was like this way or not. But I had deeper impression on the regional planning of Sentosa in Singapore. There are different theme park areas, such as Marine Museum, the Global Theatre , Casino, Commercial Street, and hotels of various levels are also available here. There are Four-star Hard Rock as well as hotels with theme of Spa, including the "Seabed" which we had stayed (Actually it was one largest Aquarium Hotel in the Ocean Park : EQUARIUS HOTEL) ,without any noisy commerce or Casino, which is a pure place for vacation. Each region is connected by B/O Shuttle Bus, thus such plan can meet different requirement of the customers. I believed the Changlong after being completed will give the tourist more surprise.

我们期待着。

I looked forward to that.

INTERCONTINENTAL.
HOTELS & RESORTS
洲际酒店及度假村

上海瑞金洲际酒店

40 ★★★★

INTERCONTINENTAL, SHANGHAI RUIJIN

Address : No. 118 Ruijin ER Road Shanghai, 200020 China, People's Republe
中国上海市瑞金二路118号上海，200020

Telephone :+86 21 64725222

Http : //www.ihg.com/

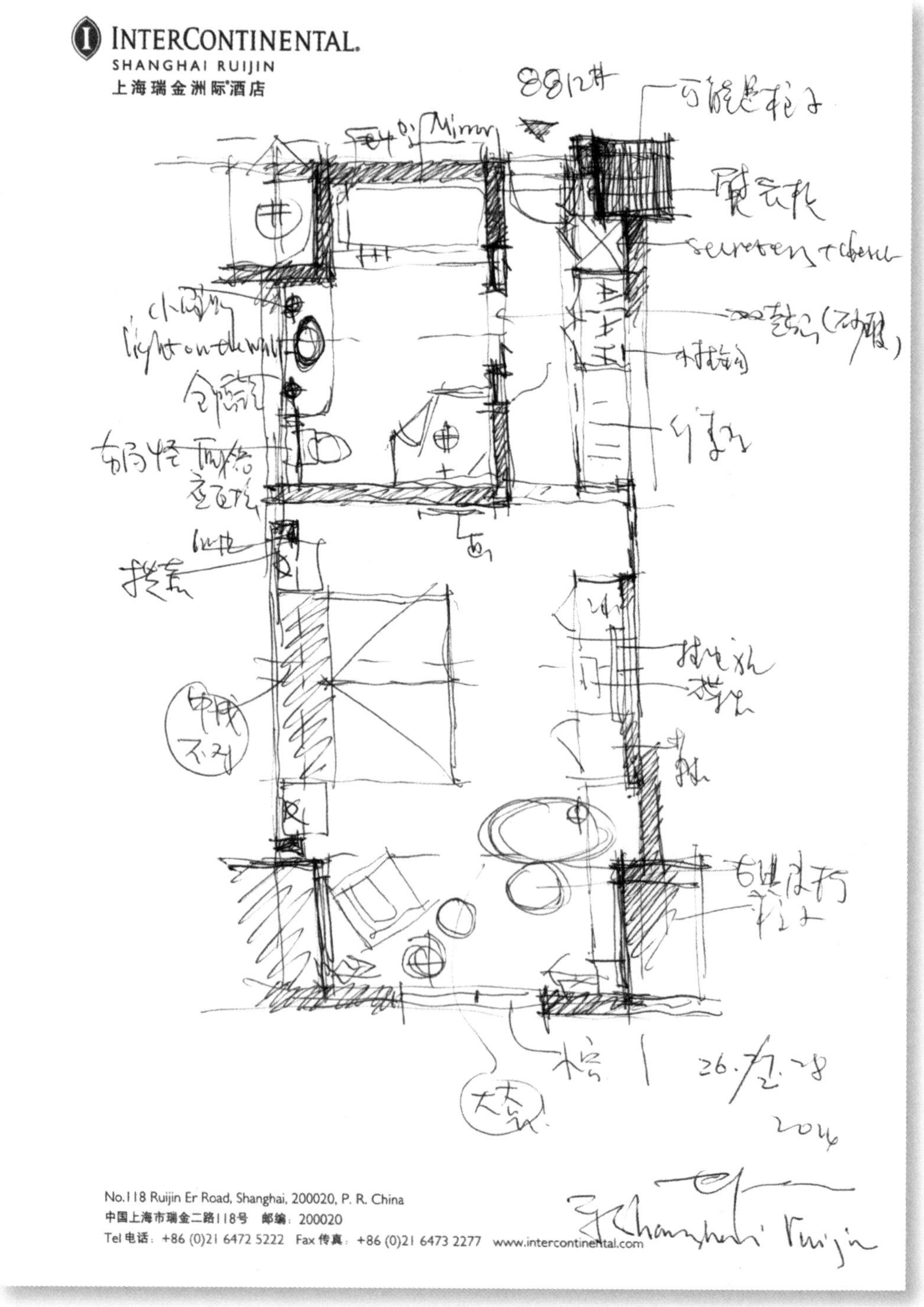

INTERCONTINENTAL.
SHANGHAI RUIJIN
上海瑞金洲际酒店
No.118 Ruijin Er Road, Shanghai, 200020, P. R. China
中国上海市瑞金二路118号 邮编：200020
Tel 电话：+86 (0)21 6472 5222 Fax 传真：+86 (0)21 6473 2277 www.intercontinental.com

INTERCONTINENTAL.
SHANGHAI RUIJIN
上海瑞金洲际酒店

"怀旧"

叶
28/2 2016

这次选了一家刚开的"新"酒店住，原来的"瑞金宾馆"现在是"洲际"了，也不知道是不是国内酒店变成的一个子品牌，还是"国外"连锁品牌管理的结果。因为没有保持原有的古老建筑风貌，而挂起了一个新招牌，不是招牌（国际）管理集团"民族的""世界的"吗！这合理吗？！

"怀旧"是它的重点，也是它沉淀下来的价值，这些好的价值，新的酒店在旧有的基础上会更珍惜吗，从这家瑞金洲际酒店可见一斑：

一、价格高企，1700元的一天客房，太高了；

二、装修很不能再普通的了，只能打60分，不说惊喜，实在失望；

三、服务员不如我们，不合格！（接待一般，）甚至有时连早餐也不用说"新"酒店味道了，很失望！待…

No.118 Ruijin Er Road, Shanghai 200020, P.R.China
中国上海市瑞金二路118号　邮编：200020
Tel 电话：+86 (0)21 6472 5222　Fax 传真：+86 (0)21 6473 2277　www.intercontinental.com

INTERCONTINENTAL.
SHANGHAI RUIJIN
上海瑞金洲际酒店

难得！没想到这么好的酒店，这么好的"故事"，这或者是现代中式（或新上海）所不属有的！

那么瑞金，同样那座公寓……有完整的建筑，也能有你在上海的贵族风。希望这些的建筑能够不错，相信你们会用新技术（追求现代的设计），可能就是酒店的设计来源。在此借这份酒店的加入会将是一个全新的中西合璧的地方，也是一种好的设计，追求"怀旧"，尽可能"旧新"再现又"原有"的新酒店，也就是个好价钱！怀旧，全都不要动！

作为入住者，不仅要把握好品质的用心的，更不会忘记"怀旧"是什么样的情调，倒不是，或者什么……之上了，更重要的是实在的历史的传承，让你有个梦中的回忆！好好"怀旧"一下！

No.118 Ruijin Er Road, Shanghai, 200020, P. R. China
中国上海市瑞金二路118号　邮编：200020
Tel 电话：+86 (0)21 6472 5222　Fax 传真：+86 (0)21 6473 2277　www.intercontinental.com

……在上海，……！

上海瑞金洲际酒店 *InterContinental, Shanghai Ruijin*

"Nostalgia"
"怀旧"

这次选了一间刚开的"新"酒店住，原来的"瑞金宾馆"现在是"洲际"了，也不知道是不是国内酒店宿命的一个写照，还是以国际连锁品牌管理为终点。国产货好像始终地守着基层，而托起了一系列老牌，不老牌的国际管理集团，"民族的"斗不过"国际的"！这会改变吗?!

This time I chose a "new hotel" which was just opened. It used to be RUI JIN Hotel but now is InterContinental Hotel. I'm not sure whether it is an example for the domestic hotel's fate: One day it will be taken over by an international hotel brand. The domestic hotels are always on the bottom class, and many international hotel management groups, no matter new or old ones, develop by taking over them. The "Foreign one" beats the "national one"! Will it change?

"怀旧"是它的卖点，也是它沉淀下来的价值，这是时间的代价，新的酒装在旧的瓶子里，会怎样呢？从这间瑞金洲际酒店可见一斑。

"Nostalgia" is the topic of this hotel, and also its property of long history. How about having new management and business in the old place? We can see in this InterContinental Shanghai Ruijin.

一、价格高，1700元一天的单人房，挺高的。

A. The price is very high. A single room cost 1700 Yuan per day.

二、普通得不能再普通的面积及布局，只能打60分，不说惊喜，应说失望。

B. Both room size and room layout are very common. I can only give 60 points. No surprise, but disappoint.

三、装修水平不如我们，投入一般，不合格！老式但没有味道，更不说“新酒”的味道了，很失望！徒有虚名！浪费了这么好的地段，这么好的“前身”。这或者是前任中方（或物业主）所不愿看到的！

C. Interior of room is worse than our works. The investment is not enough. Not good! It's nostalgic without individual style or even new feeling. It made me disappointed! That belied it name! What a waste for such a good location and good construction! I think its former Chinese host (or owner) don't want to see that!

上海瑞金洲际酒店 *InterContinental, Shanghai Ruijin*

平面疏松，园林环境有致，有高低错落，建筑物有老上海的骨风。新改造的主楼空间不错，标准层实用率很低（追求形式的结果）。可能就洲际的品牌来说，在上海这间酒店的加入会拉升其在一线金融中心的地位，也是一种好的方式，追求“怀旧”。尽收“旧瓶”再灌入“低劣”的新酒，也能卖个好价钱！怀旧，有时不容易！

The layout is common, but the garden is beautiful. Buildings are scattered low and high, with the style of old Shanghai. The space of main building after reconstruction is nice, but efficiency rate of standard floors is low because designer concerned to appearance more than interior. Maybe according to InterContinental brand, taking over Ruijin Hotel Shanghai can upgrade the image for the whole district. It's a nice way to be "nostalgic", although such new design is not very suitable for old construction, the price is still good! Sometimes it's even more difficult to be nostalgic!

作为入住者，不住过是体验不到其经营者的用心的，更不会感受“怀旧”是什么样的结果。住了两天，或者你就慢慢喜欢上了。更主要的是它演绎出对历史的传承，让你有参与其中的经历！好好“怀旧”一下！特别是在老上海，老区里的！

As a customer, I can't understand the idea of business operator, or even can't feel the "nostalgia" before I stay in the hotel. After staying for two days, maybe you will like it. What is more important is that it shows you how to inherit history and makes you feel each detail. Just come for nostalgia! Especially if you are from old district of Shanghai!

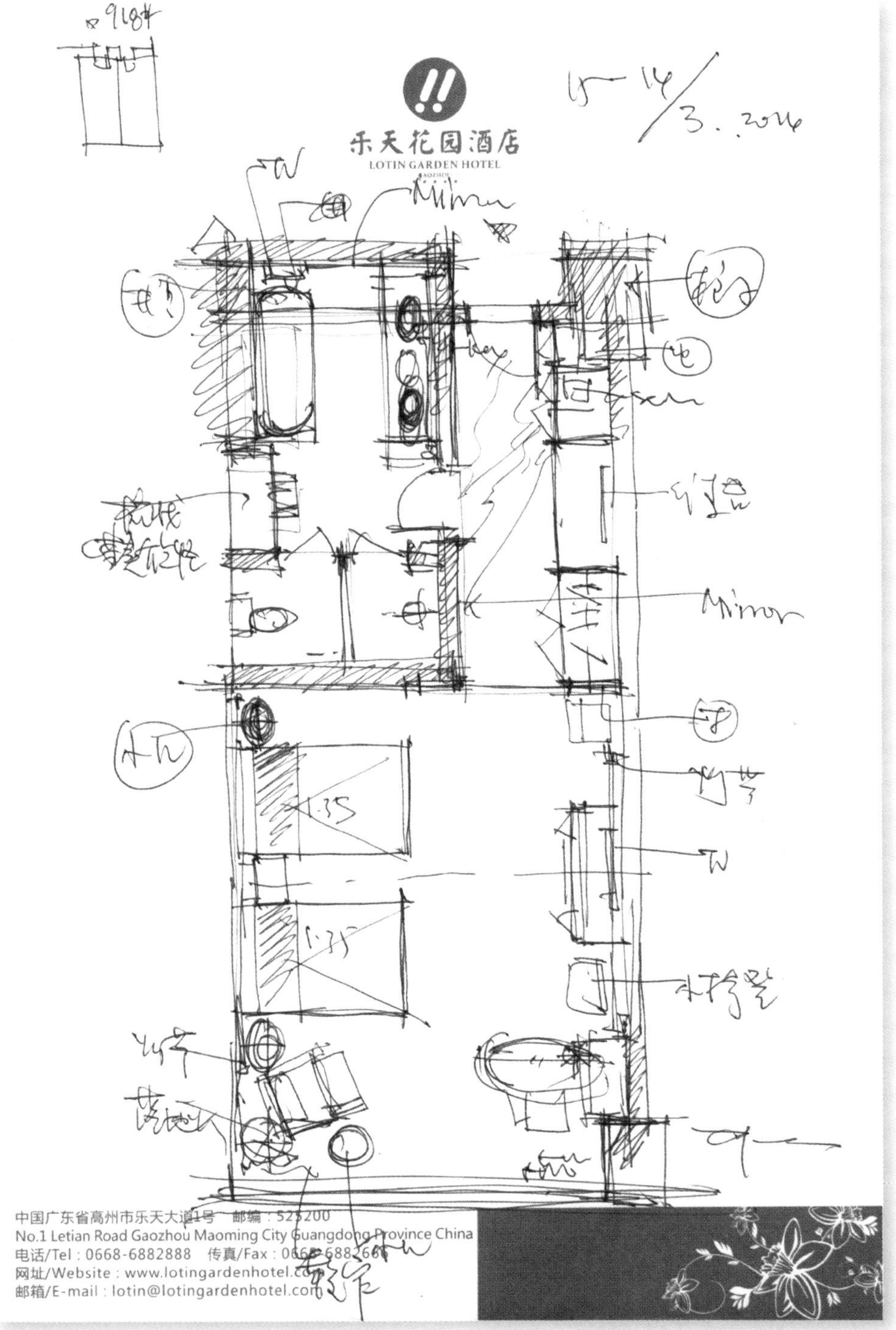
乐天花园酒店
LOTIN GARDEN HOTEL
Mirror
Mirror
1.35
1.35
中国广东省高州市乐天大道1号　邮编：525200
No.1 Letian Road Gaozhou Maoming City Guangdong Province China
电话/Tel：0668-6882888　传真/Fax：0668-6882666
网址/Website：www.lotingardenhotel.com
邮箱/E-mail：lotin@lotingardenhotel.com

乐天花园酒店
LOTIN GARDEN HOTEL
高州乐天花园酒店
★★★★★

41

LOTIN GARDEN HOTEL, GAOZHOU

Address : No.1 Letian Road Gaozhou Maoming City Guangdong Province China
中国广东省高州市乐天大道1号
Telephone : +86 0668 6882888
Fax : +86 0668 6882666
E-mail : lotin@lotingardenhotel.com
Http : //www.lotingardenhotel.com /

写于酒店 16/3.2016

意外的收获

意外的收获而节省幸之！同学在介绍的项目之要之“乐天”的酒店是值得来高州的一个好地方，项目经理是我们的一位师弟同门之弟的一个酒店项目，“以前”我学习的多年专业，也是治我和我的同学共同合作之建筑与室内专业一线实践！正是酒店建设的最好时期，再去到此地讨论一下乐天酒店的优势与我们在广西岑溪酒店的现状与分析，当然也结合当地的文化、市场现状，探讨策略而对酒店项目的功能的需求和开发周期会增加合理，提出了我们的初步建议，它们是一步步建立的，在于会与意外之基础上，我们将是最重要的，相信成功的酒店也一样，意外与收获是一种开始，不能忘记！

Accident and Coincidence
意外的巧合

意外、巧合而带有幸运！同学在广西酒店项目的业主“看”的酒店是位于广东高州的一间。打听一下，巧得很，是我们的一位帅哥回公司之前一个酒店项目，“顺理成章”得以多重考量，业主让我和我的同学共同合作：建筑与室内专业一气呵成！正是酒店建设的最好的方式。再次到现场讨论一下乐天酒店的优势与我们在广西岑溪酒店的现状与分析，当然也结合当地的文化、市场现状、投资策略而对酒店项目的功能配置和开放周期合理化分配，提出了我们的初步建议。信任是一步步建立的，在巧合与意外之基础上，我们的努力是最重要的。相信成功的酒店也一样，意外与巧合只能是一种开始，不能长久！

Accident and coincidence brought fortune to me! My classmate had a hotel project in Guangxi Province, and his client and we went for investigation to a hotel in Gaozhou City, Guangdong. After inquiring about the hotel, I found it's coincident that it was designed by one of our designers before he joined my company. Because of that, after consideration, the client asked my classmate and I for cooperation: He works for architechture design, and I work for interior design! This is the best way to build a hotel. In the hotel, we talked about the advantage of Lotin Hotel and current status analysis of Genxi Hotel Guangxi. We gave our primary advice about the functional arrangement and rational opening time planning of the hotel, according to the local culture, marketing condition and investing tactics. Trust is built step by step. Base on accident and coincidence, our effect is the most important. I believe it's the same for a successful hotel. It started from accident and coincidence, but can't last for long.

高州乐天花园酒店 *Lotin Garden Hotel , Gaozhou*

WALDORF ASTORIA®
北京华尔道夫酒店
★★★★★

42

WALDORF ASTORIA, BEIJING

Address : No.5-15 Jinyu Hutong Wangfujing Dongcheng District,Beijing,100006,China
中国北京市东城区
金鱼胡同5–15号，100006

Telephone :+86(0)10 8520 8989

Fax :+86(0)10 8520 8990

Http : //www.waldorfastoria.com/beijing

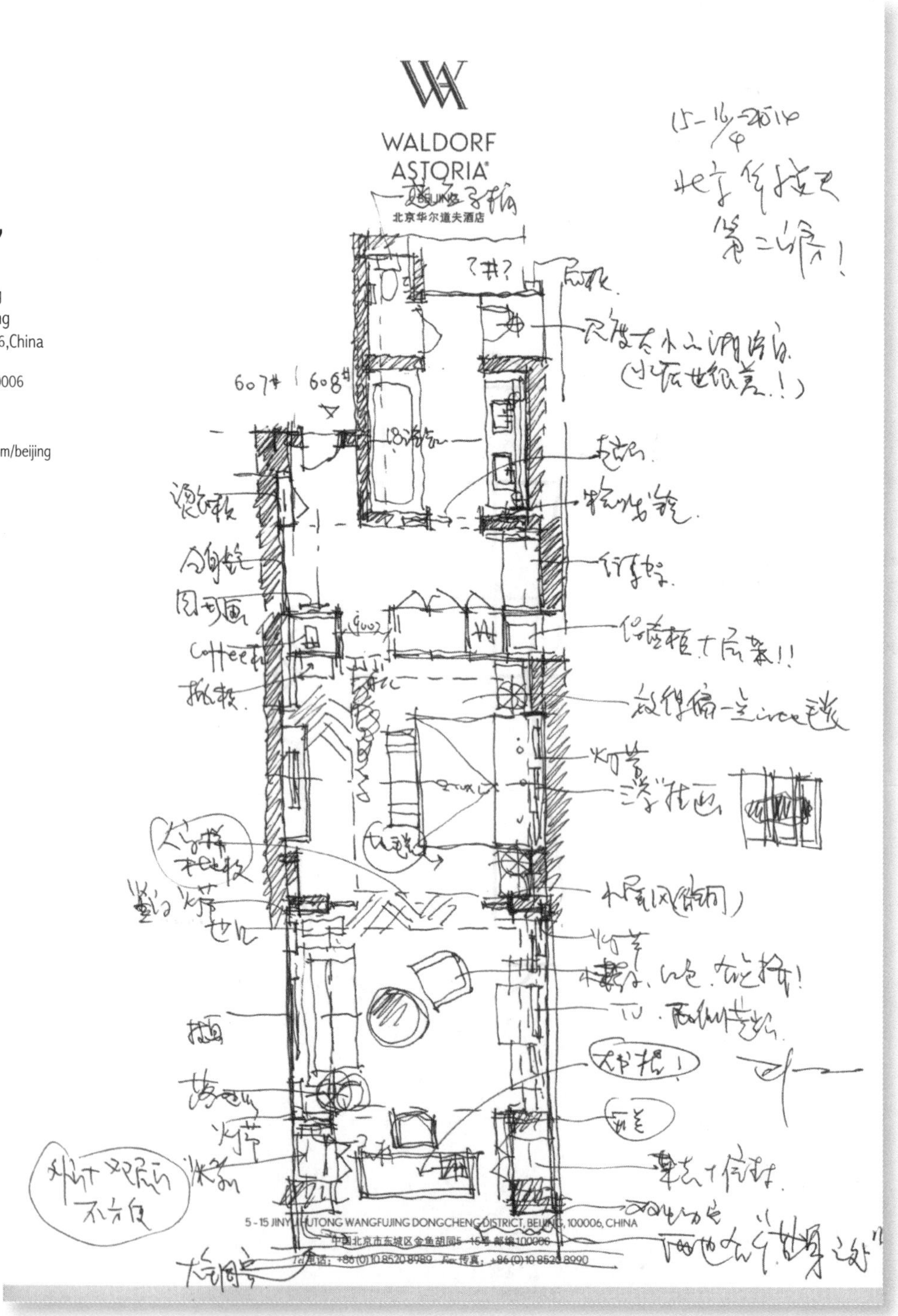

WALDORF
ASTORIA®
BEIJING
北京华尔道夫酒店
607#
608#
5-15 JINYU HUTONG WANGFUJING DONGCHENG DISTRICT, BEIJING, 100006, CHINA

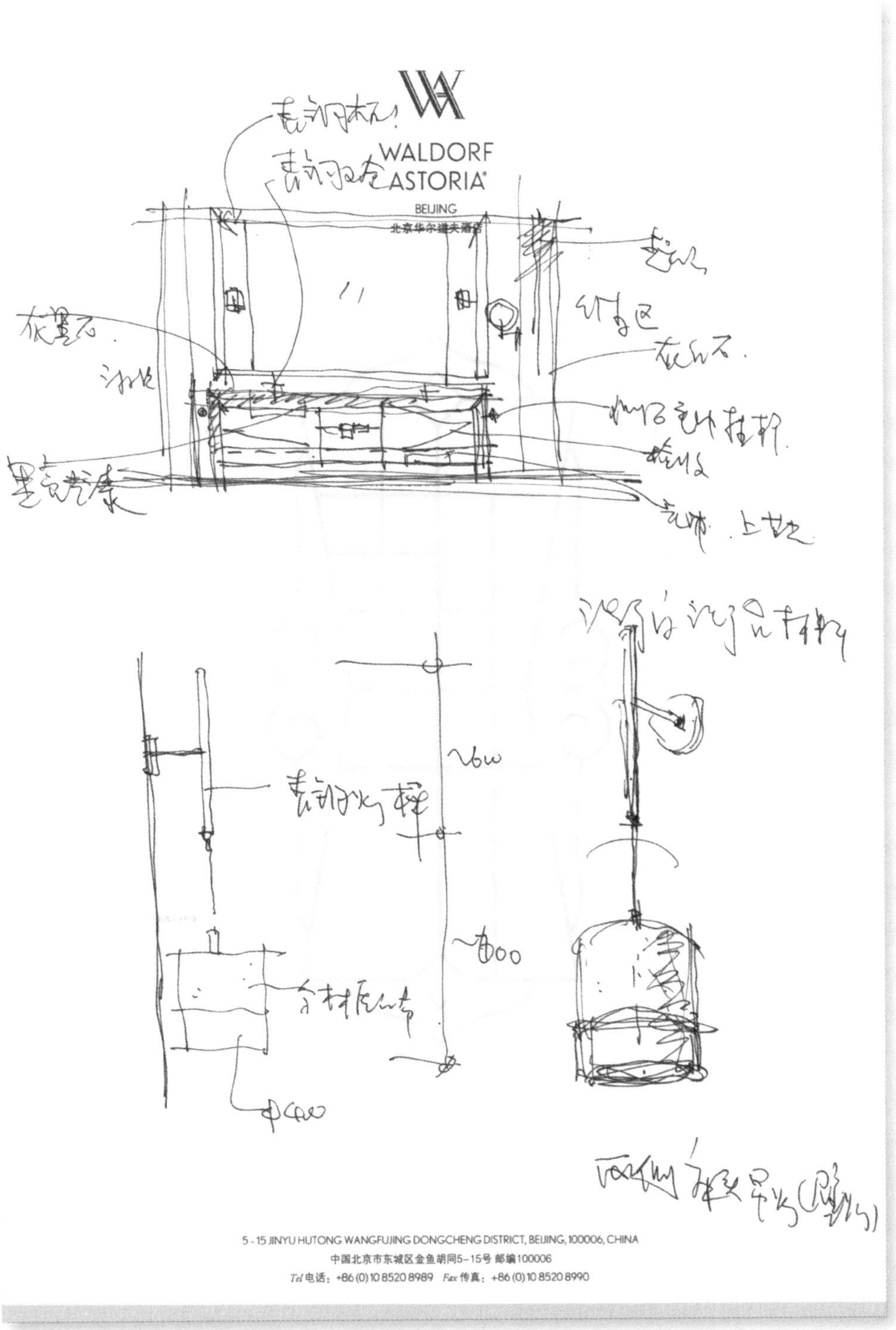
WALDORF
ASTORIA®
BEIJING
北京华尔道夫酒店
5 - 15 JINYU HUTONG WANGFUJING DONGCHENG DISTRICT, BEIJING, 100006, CHINA
中国北京市东城区金鱼胡同5–15号 邮编100006
Tel 电话：+86 (0) 10 8520 8989 Fax 传真：+86 (0) 10 8520 8990

WALDORF
ASTORIA
BEIJING
北京华尔道夫酒店
5-15 JINYU HUTONG WANGFUJING DONGCHENG DISTRICT, BEIJING, 100006, CHINA
中国北京市东城区金鱼胡同5-15号 邮编100006
Tel 电话：+86 (0) 10 8520 8989 Fax 传真：+86 (0) 10 8520 8990

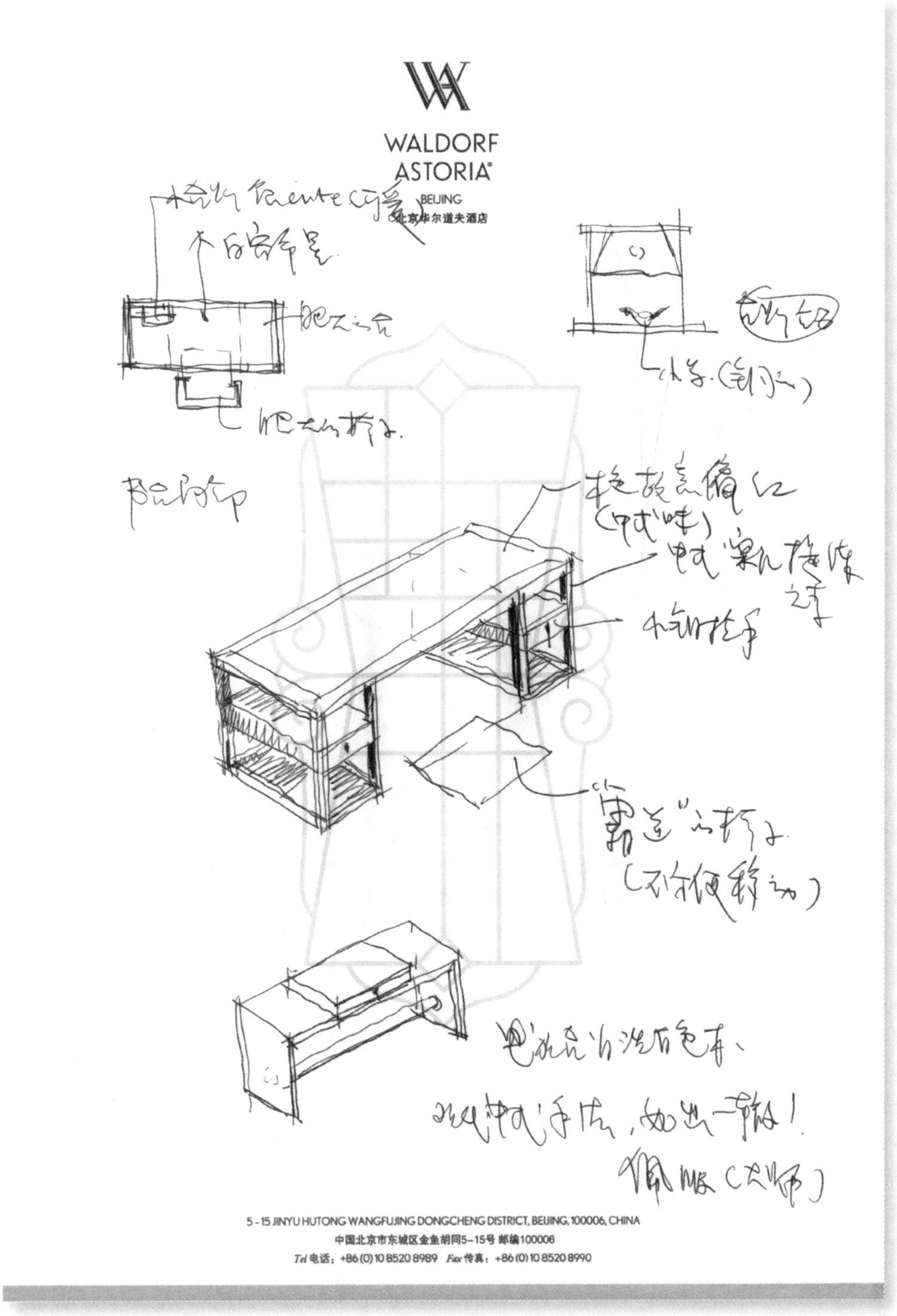
WALDORF
ASTORIA®
BEIJING
北京华尔道夫酒店
5 - 15 JINYU HUTONG WANGFUJING DONGCHENG DISTRICT, BEIJING, 100006, CHINA
中国北京市东城区金鱼胡同5–15号 邮编100006
Tel 电话：+86 (0) 10 8520 8989 Fax 传真：+86 (0) 10 8520 8990

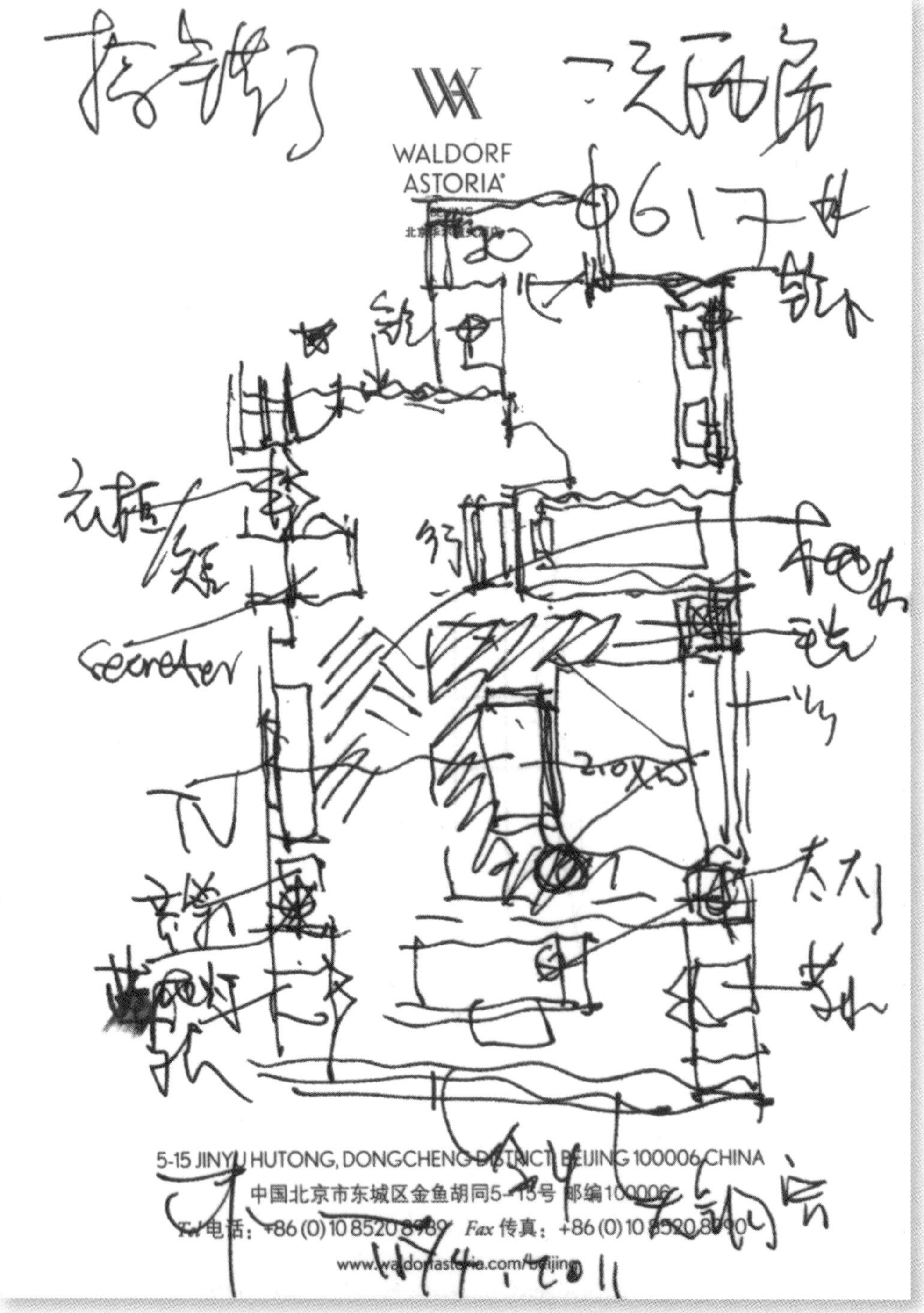
WALDORF
ASTORIA
secreter
TV
5-15 JINYU HUTONG, DONGCHENG DISTRICT, BEIJING 100006 CHINA
中国北京市东城区金鱼胡同5-15号 邮编100006
www.waldorfastoria.com/beijing

且行且完善

印象中只有非五星级酒店会边开业边逐步完善配套的一些功能和设施。而近年发觉入住的一些新开业的五星酒店有不少是不完整开业的，大多数是娱乐（可能未找到经营者或未装修好吧）康体部分的不完善（主要是泳池区）。如长沙的北辰洲际酒店新开业泳池也还未建好，文华东方（上海）酒店也是。长隆集团在珠海的横琴湾酒店更甚，只完成了小部分客房，自助餐厅、食街就匆匆忙忙投入使用。不少空的是周边设施，但更有早上的噪声，周边的工地，四处的异味，当然赶上了些许的黄金档期，据说去年时（2014年）还订不到房间（饥饿营销），利益驱使的结果，我们都成了吸声、吸味、吸尘的小白鼠了。

上海瑞金洲际酒店的开业会特殊些，因为是旧酒店收回改造加建主楼，新旧并融，市中心的大大占地，经典怀旧式的花园，大幢各个独栋别墅式客房都是其很大的优势，入住时带spa优惠券去体验一下熟，这些区域：室内恒温泳池、健身、spa 还是旧时的样子，经营状况良好。（1985年开业，2004年翻新）部分别墅式客房已改造过了，部分还在装修，依旧的新建主楼很好地与环境相一致。说明之开业，但也相当令人满意的。

这里大师跟了好几年的王府井全面翻修的半岛酒店也一样，开业两年后泳池区才投入使用，而招牌的四合院客房还要再伸长脖子耐心等一下（装修时间）。

时间能养好东西，好东西也需时间慢慢养的，时间是最大的价值，故曰"自行自完善"吧！

北京华尔道夫酒店 *Waldorf Astoria, Beijing*

Good and Perfect
且行且完善

印象中只有非五星级酒店会边开业边逐步完善配套的一些功能和设施，而近年发觉入住的一些新开业的五星级酒店有不少是不完整开业的，大多数是娱乐（可能未找到经营方式或未获得牌照）、康体部分的不完整（主要是泳池区）。如长沙的万达文华酒店新开业泳池区未完工，文华东方（上海）酒店也是。长隆集团在珠海的横琴湾酒店更过分，只完成了小部分客房、自助餐厅、食街就匆匆忙忙投入使用。不少空间是围蔽施工，但更有早上的噪声、周边的工地、四处的异味。当然赶上了过年前的黄金档期，据说过年时（2014年）还订不到房间（饥饿营销），利益驱使的结果，我们都成了吸声，吸味、吸尘的小白鼠了。

It seems that only the ones who is not Five-star Hotel will improve some of their functions and facilities gradually since opening. However, we find that very few Five-star Hotel run its core business or we can say it is incomplete opening. Most of them are for recreation (maybe it can not find the right way of business model or can not get the license), incomplete healthy SPA service (main part is focused on the swimming pool area). For example, like Wanda Vista Changsha still fails to complete the construction of its new opening swimming pool area, and Mandarin Oriental (Shanghai) Hotel is the same. Hengqin Bay Hotel which is owned by Changlong Group in Zhuhai even too much and it only finished a small portion of guestroom ,cafeteria,food-street. It was constructed in besieged area, but it was too noisy in the surrounding construction site in the morning and sent out smelly smell.Of course, when it was speeding construction so that it could catch the golden period before the Spring Festival. It's said that near the Spring Festival of 2014, customers still could not book the room (Maybe it is called Hungry Marketing).The result that the benefit had driven, we have become the so-called little white mouse to feel the noisy sound, inhaling the disgusting smell and dust.

上海瑞金洲际酒店的形式会特殊些。因为是旧酒店收购改造加建主楼，新旧兼容，市中心的大大占地、经典怀旧式的花园、大幢而分散的别墅式客房都是其强大的优势。入住时带spa优惠券，去体验一下看看。这些区域：室内恒温泳池、健身、spa还是旧旧的样子，经营有好些日子了（1985年开业，2004年翻新）。部分别墅式客房已改造迎宾，部分还在装修。仿旧的新建主楼很好地与环境相一致。说匆匆开业，但也是相当令人满意的。

The form of Ruijin InterContinental Shanghai will be more unique. Due to the old hotel acquisition and transformation and the on-building main building, the fusion of the old and the new, the central downtown area occupied quite a lot of land, the old classic and nostalgic garden, big-block but scattered villa types of guestroom are its strong advantage. You can bring your spa coupons and go to experience when living in, these areas including indoor heated swimming pool, keep-fit gym, spa are still the old appearance as it has been running (since 1985 and re-decorated in 2004). Some of the villa type of guestroom has been re-built and some are still under decorating. The new main building which is built by imitating the old one can be well integrated with the environment. Although it is opened hastily, however, all in all, it is quite satisfactory.

乃至大师跟了好几年的王府井金鱼胡同的华尔道夫酒店也一样，开业两个月泳池区才投入使用，而招牌的四合院套房还要再伸长脖子耐心等一下（未公报时间）。

Even the Waldorf Astoria Hotel which located in Wangfujing Goldfish Alley that the Great Master had been following carefully for several years is the same. Only when the swimming pool is opened for 2 months can it be put into use. However, for the characteristic courtyard suite, I am afraid we have to wait for some more days (as the time is still not decided yet). So, Just be patient.

时间能养好东西，好东西也要时间慢慢养。时间是最大的价值，权当“且行且完善”吧！

Time can make good things, and good things also need to take time to be perfect. Time is the biggest value, so we give it the name "Good and perfect", which means it takes time to be good and perfect!

铜活

有报告，本人住已从各种渠道看到了它的开张——伯的北京华尔道夫酒店的开启，Tabu在中国，特别是在首都之大作是一个"真铜"的项目。到此现场，相当震撼，外表"铜装"包裹相当华丽。但细看之下，"多种铜态"同时，而且这种不统一一直治（延）续到整个酒店，包括客房的细节！

印象中，铜还是相当易氧化的，会发铜绿，以前在项目中是取少量和小范围使用，或者此代的工艺技术或通过合金的方式、表面处理的方式达到更好的耐候性，但事实上在运用上还是慎重的。

华尔道夫，外立面的窗框可能是铜合金，发色处理，还是相当的亮丽，而大面的铜板外墙已经发雷发黑，或是大师故意为之，或不得已，入口的大门框的板材加工还是相对保障一些，包括公共区域的铜收边线。铜板（薄板）切割的漏花屏风还是相当好的，有风采。可能线条化的不停步看样病吧！

床的黄铜部分的还是太新、王金、极致高端、完美"私人订制"，出于加工要求，可能相对偏轻、偏小（合金原因），入住使用以后发现上面有掉不去的斑斑点点水渍，看来大师的"铜活"不易讲出。

但凡是不沾水的配件：拉手、合页、毛巾挂杆、行李架防撞杆、壁灯的固定吊/挂杆都是全铜订制，且作为套装完成，每一处都历含心思，但可能不是出于同一批的同一厂家，表面处理也不同，所以深浅、粗细、光泽都有不同，看来需请一试那位大师了。

或者我们可以与时间同步，任之与铜一齐氧化，静静地延续大师的"铜活"。

北京华尔道夫酒店 *Waldorf Astoria , Beijing*

有预告，未入住已从各种渠道看到了新开张一个月的北京华尔道夫酒店的芳容。Yabu 在中国，特别是在首都之大作是一个“卖铜”的项目。到达现场，相当震撼，外表“铜装”包裹相当华丽。但细看之下，各种“铜态”同现，而且这种不统一一直沿（延）续到整个酒店，包括客房的细节！

From the pre-advertisement and other channel, we have seen the appearance of Waldorf Astoria in Beijing which is just opened for one month although we never stayed there. The big project of Yabu in China, especially in the capital is the "Copper selling" project. When we arrived at the site, we are quite shocked. The outer appearance is wrapped with "copper clothes", which is indeed splendid and extravagant. When we see it carefully, you will find various types of "copper shape" appearing and this seemed never unified with the scene, however it is extended to the entire hotel including the details of the guestroom!

Copper Story
铜话

印象中，铜还是极易氧化的，会发铜绿，以前在项目中只敢少量和小范围使用。或者现代的工艺技术或通过合金的方式、表面处理的方式使其达到更好的耐候性，但事实上在运用上还是较慎重的。

In our impression, the copper is extremely easy to be oxidized and will become bronze green. We seldom or only used a little bit in our project before. Although the modern art and workmanship or the way of metallic alloy has greatly been improved to achieve the durability through the surface processing, as a matter of fact, we are still very cautious when putting into use.

华尔道夫，外立面的窗框可能是铜合金发纹处理，还是相当的亮堂，而大面的铜板外墙已经发霉发黑，或是大师故意为之，或不得已。入口的大门框的板材加工还是相对有保障一些，包括公共区域的铜收口线，铜板（薄板）切割的镂花屏风还是相当的有风采。可能线条化的不容易看出毛病吧！

Waldorf, the window frame of the outer surface may be processed by bronze metallic thread, and it is still quite bright. However, the outer wall of the bronze plate with big surface has already been moldy or black. I don't know if the Building Master makes it on purpose or not. Or the Master has no choice but to choose this. The wood processing of the big door frame is relatively secure including closing line of the copper in public area, the screen stencil that cut from the bronze plate (laminated sheet) is still quite elegant. Maybe the threaded one is not easy to find the defect!

这个建筑给我的最深印象的还是龙头、五金，是极致高端、完美的“私人订制”。出于加工要求，可能相对偏软、偏红（合金原因）。入住使用时已发现上面有抹不去的斑斑水渍，看来大师的“铜话”是不易讲的。

The deepest impression of the house leaves me is still the faucet, hardware, extremely high-end, perfect "private-tailored". It may be slightly soft as it is out of processing requirement. It is reddish (maybe it is caused by the metallic alloy). We have found some stained water which can not be erased when living there. See from this, it seems that the Master's "copper job" can not be commented in just one or two words.

倒是不沾水的配件：拉手、合页、毛巾挂杆、行李架防撞杆、壁灯的固定吊/挂杆都是全铜订制，但依然柔美亮丽，每一处都历尽心思，但可能不是出于同一品牌、同一厂家，表面处理也不同，所以深浅、粗细、光泽都有不同。看来要统一就可难为大师了。

Parts that never stained with water like handle, hinge, tower hanging bar, luggage bumper, fixing hanger for the wall lamp, which are all made from copper. It is still soft and beautiful. Everywhere you see is quite different and particular, however, they may not be made from the same factory or the same brand. The surface processing workmanship is quite different as well in terms of deep and light color, thickness or thinness, glossy, etc. It seems that it will be a touchy job for the Master to unify them together.

或者我们可以与时间同步，慢慢与“铜”一起氧化，静静地延续大师的“铜话”。

Or we can keep abreast with the time and slowly "oxidized" like the copper and wait patiently for continuing the masterpiece copper story made by the Master further!

小故事，小事故

这种事情，确实未发生过。

一天在同一酒店内住"了两间房，值得纪念。

按预订房、直奔入住，没想过会出错，人的思维就是这样，因为信任，就没有细究了。

第一间房是正常的标准大床房，稍偏宽，入房后随手用手机拍下来作记录，同时翻翻箱，倒倒柜，算是带着兴趣来参观新开的Yabu先生的作品[illegible]酒店，惊喜不少：配色、用材、家具等等，真是深厚功力！

及晚上与同行聊起这间酒店的不同房型，才知道我是“故意”订了附客厅的房型，于是匆匆去“质问”前台，哈哈，好似维权行动，原来是酒店出了小事故，给错了房型，便匆匆马上在午夜再匆匆地搬进到我应得的长长条状的房间，当然赔来了一个大果盘和一连串的道歉，也算是第一次住过的“大土豪”金鱼胡同“铜”酒店的精彩遭遇了。

一天住两间房，一宽一长型，与上次在兰州皇冠假日酒店时
两个大男人先入住大床房再转到标准双床房的经历有
异曲同工之乐趣！
不管怎样，让我便宜了一间。
有时候，小事故，成就了小故事。

Small Accident Makes a Story
小故事，小事故

这种事情，确实未发生过。

For this case, actually it never happened in my life.

一天在同一酒店内“住”了两间房，值得纪念。

Live in two rooms in the same hotel in one day, it is one memorable day.

携程订房，正常入住，没想过会出错。人的思维就是这样，因为信任，就没有细究了。

Booking from Ctrip and live as usual, I never think it will have mistake. People always think it that way. Due to the trust, we are not so particular.

第一间房是正常的标准大床房， 稍偏宽。入房后随手用手机拍拍来作记录。同时翻翻箱，倒倒柜，算是带着兴趣来看看新开的Yabu先生的华尔道夫酒店。惊喜不少：配色、用材、家具等等，颇具深厚功力！

The first room is the very normal standard room with big bed, slightly wider. Take picture by mobile and make some recording after checking in. While turning over the box and checking the cabinet, we are here to see the newly-open Waldorf Astoria Hotel which is introduced by Mr. Yabu with strong interest. It indeed brings you surprise: Color-matching, material using, furniture, etc. All these are so particular and careful.

北京华尔道夫酒店 *Waldorf Astoria , Beijing*

及晚上与同行聊起这间酒店的不同房型，才恍然，我是“故意”订了有小客厅的房型，于是匆匆去“质问”前台。哈哈，少有的维权行动。原来是酒店出了小事故，给错了房型，便匆匆画下了这平面，再匆匆地搬迁到我应得的长长条状的房间里。当然赚来了一个大果盘和一连串的道歉，也算是第一次住这儿的“大土豪”金鱼胡同“铜”酒店的精彩遭遇了。

Chat with my peer in the evening about the different types of rooms in this hotel. Only at that time can we realize that I book one small type of room with meeting room, and then I go hastily to “question” the reception. Aha, the very few time that people will complain. Consequently the hotel made some mistake, and they gave me the wrong type of house,so I hastily drew this plane of layout, and then moved hastily to the room with long strip that I had booked before. Of course I gained a big fruit plate and apology as repay. That is my wonderful experience in my life and the first time to enjoy living in this “luxurious” “copper” hotel in the golden fish alley.

北京华尔道夫酒店 *Waldorf Astoria , Beijing*

一天住两间房，一宽一长型，与上次在兰州皇冠假日酒店时两个大男人先入住大床房，再转到标准双人房的经历有异曲同工的乐趣！

I have lived two rooms in a day. One is wider and one is longer type. Compared with last time when we two guys living in a room with big bed in Crown Plaza Lanzhou and then transfer to standard double room, I think it has the same fun and pleasure!

不管怎样，让我占便宜了。

Anyway, it is really a good bargain that I have got.

有时候，小事故，成就了小故事。

Sometimes, one small accident will make it a memorable story.

éclat

BEIJING • 北京

北京怡亭酒店

43

★★★★★

HOTEL ECLAT, BEIJING

Address : No. 9, DongDaQiao Road, Parkview Green FangCaoDi , Chaoyang Dist., Beijing 100020
北京朝阳区东大桥路9号
侨福芳草地 100020

Telephone :+86 10 8561 2888

E-mail : beijing@eclathotels.com

Http : //www.eclathotels.com/

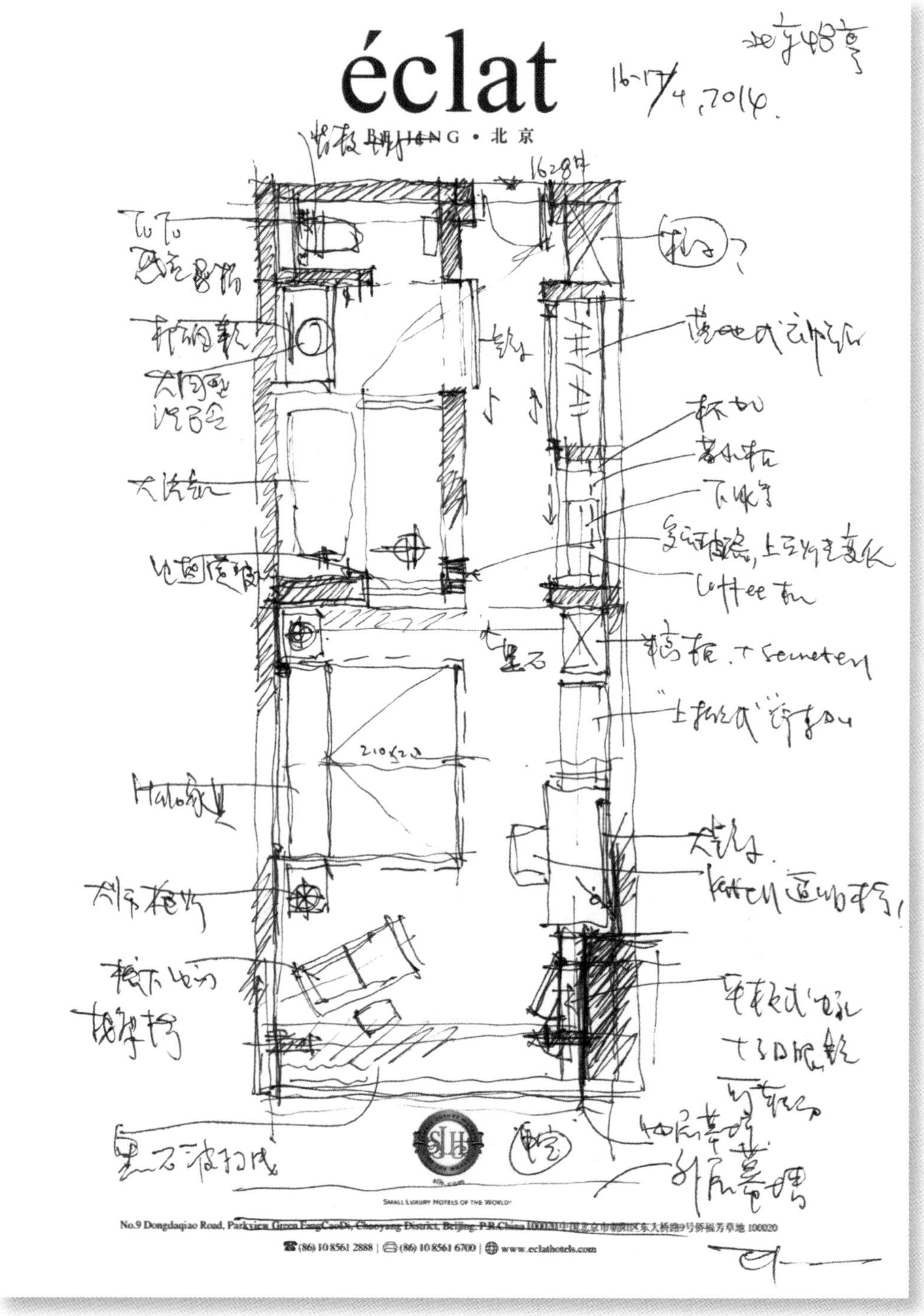
éclat
BEIJING • 北京
210x220
SMALL LUXURY HOTELS OF THE WORLD
No.9 Dongdaqiao Road, Parkview Green FangCaoDi, Chaoyang District, Beijing. P.R.China 100020 中国北京市朝阳区东大桥路9号侨福芳草地 100020
(86) 10 8561 2888 | (86) 10 8561 6700 | www.eclathotels.com

熬艺术

号称艺术的酒店真不少，早期的[illegible]，京城的逸林希尔顿，台湾的寒舍，香港的一些[illegible]……等等，我早听说也早看过，但还没有住过，这次下定决心选了北京的侨福芳草地怡亨酒店，可谓玻璃艺术的一片草地！

入口之"偏"，连出租车司机都几乎要拖我的手机打出定位，问了几次酒店都说不清，停在到眼前也找不到门！

酒店在一个综合体的顶层，十六到二十一层，负一至四层是一个大型艺术装置了的商场，对的是玻璃艺术的商场，大型的、中型的，再到它结构材料构件，把整个侧玻璃围墙，一切一切，让你目不暇接，不艺术真不罢休，其中有相当的中国艺术家之大成："牛气冲天"、"玻璃的钢"、"开了天窗"，当然酒店的才是精华。在酒店首层、二层的走道和入口廊道陈展"黄先生"的收藏品，有达利的、岳敏君、方力钧、曾梵志等的大作。

可拓开眼界。

直至房间，一屋之"收藏"：Halo之门刹家具，英伦香味兰花衣领feel，木纹下多功能雕塑椅，3D影像电视机，B&O音响影院，顶棚投影式电子钟，手摇灯，当然还有最后钱之床头当代画家之大作……

结果，太艺术，均夜难眠，让人不安！

熬之呗。

Endure the Art

熬艺术

北京怡亨酒店 *Hotel Eclat, Beijing*

号称艺术的酒店真不少，早期的珠海中邦、京城的逸林希尔顿、长沙的星栈、香港的一些品牌等等。有些听过，有些看过，但还没有住过。这次下定"决心"选了北京的侨福芳草地怡亨酒店，可谓极端艺术的一片草地！

Hotel that taking art as the characteristic is never less. From the early period of Jobon in Zhuhai , Yiling Hilton in Beijing to Star Hotel in Changsha and some of the other Hongkong brands, and so on. Some I had heard before and some I have never lived though I had seen before. This time I have decided to choose one hotel called Hotel Eclat, which is located in Qiaofu grassland in Beijing, where can be called as a vast grassland full of art!

入口之"偏"，连出租车司机都几乎想把我的手机扔出窗外。问了几次酒店都说不清，待到眼前也找不到门！

It is quite "far-off and remote" place that almost drives the driver crazy. The driver has to keep asking people nearby where is the hotel, however, still no one know it. Even the hotel before you and you still can not find the door!

北京怡亨酒店 *Hotel Eclat ,Beijing*

酒店在一个综合体的顶层，十六到二十一层，负一至四层是一个大型艺术装置了的商场。对的，是被“艺术化”的商场。大型的、中型的，再到包括标识指示、扶梯侧玻璃图案，一切一切，让你目不暇接，不艺术誓不罢休。其中有相当的中国艺术家之大成：“牛气冲天”、“被锯的锯”、“开弓无箭”，当然浓缩的才是精华。在酒店首层、二层的业主私人画廊更陈展着“牛”的收藏品，有熟悉的岳敏君、方力钧、曾梵志等的大作，可谓开眼界。

Hotel is at the top floor of one comprehensive building, from floor 16 to 21. Minus 1 to fourth floor is one large-scale shopping mall decorated with art. Aha, yes, it is the shopping mall that "being decorated with art" , from large-scale to medium-scale and then the logo indication, logo on the side glass of escalator, all in all, keep catching your eyeball all the time and you will find arts are everywhere. Among of them, many master pieces are made by the Chinese Artist, such as "Bullish", "Saw being cut", "Bow without arrow", of course the condensed one is the essence. On the first and the second floor of private-own Gallery Exhibition, "Bull" collections are displayed there. There are some masterpieces from the Master Yue Minjun, Fang Lijun, Zeng Fanzhi that you may be so familiar with. It extremely broadens my eyesight.

直至房间，一屋的“收藏”：Halo的订制家具，英伦重味道，旅行家feel，松下多功能按摩椅，3D影像电视机，B&O音响系统，顶棚投影式电子钟，手枪灯，当然还有最值钱的床头当代画家的大作……

The collections are fulfilled the house: Halo tailored furniture, heavy taste of England, the feeling of the traveler, multi-functional Panasonic massage chair, 3D video TV, B&O hi-fi system, ceiling projection type of electronic bell, pistol lamp, of course, there are some most valuable masterpieces made by the temporary painter...

结果，太艺术，彻夜难眠，让你不安！

As a result, too much art may make you never easy to fall sleep and make you uneasy!

熬熬呗！

Just endure the art !

Kempinski Hotel Guiyang
CHINA

贵阳凯宾斯基酒店

44 ★★★★★

KEMPINSKI HOTEL, GUIYANG

Address : Huguo Road 68,550002 Guiyang
贵阳护国路 68号,550002
Telephone : +86 851 5999999
Http : //www.kempinski.com/en/guiyang/hotel-guiyang/

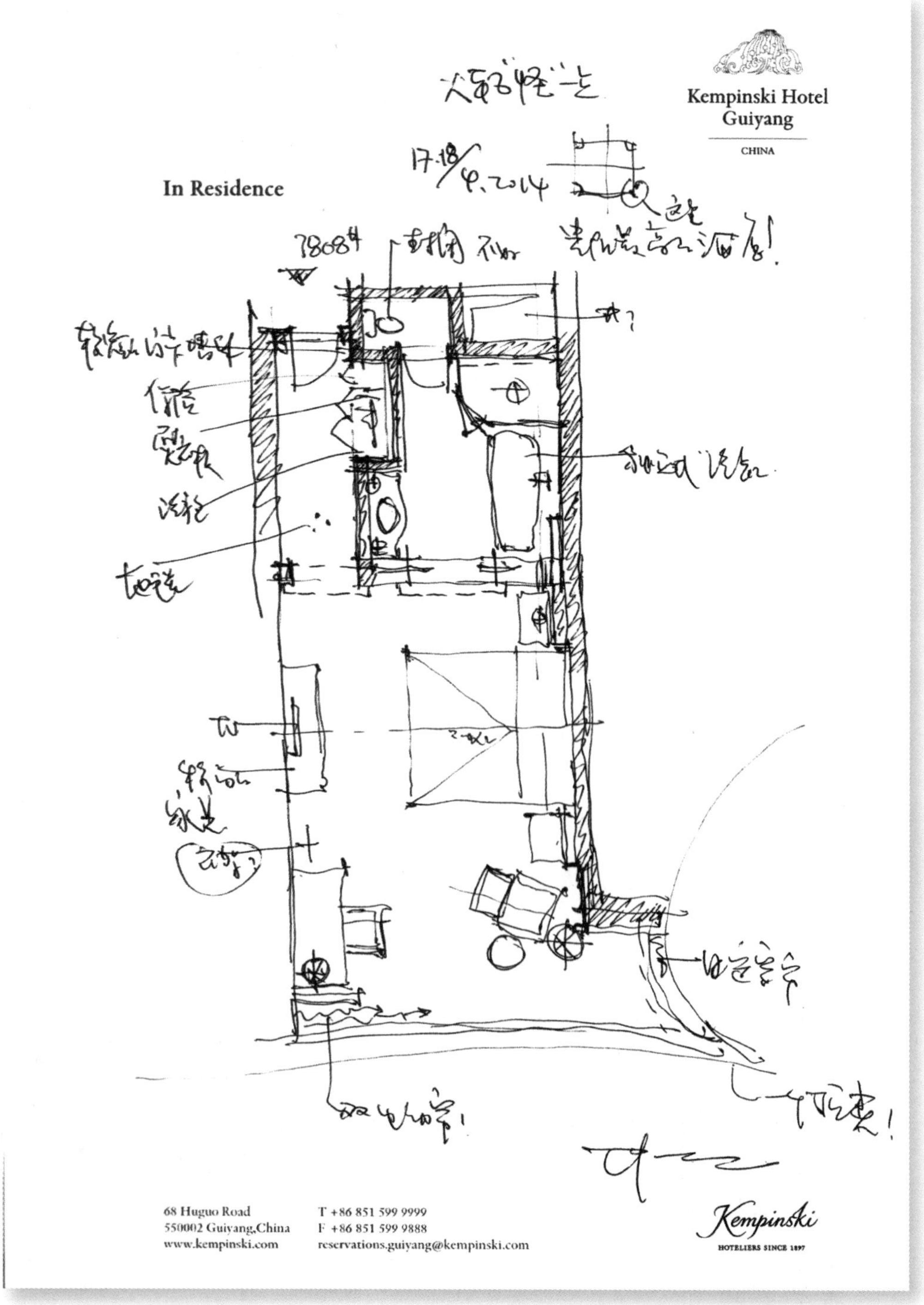
Kempinski Hotel
Guiyang
CHINA
In Residence
68 Huguo Road
550002 Guiyang,China
www.kempinski.com
T +86 851 599 9999
F +86 851 599 9888
reservations.guiyang@kempinski.com
Kempinski
HOTELIERS SINCE 1897

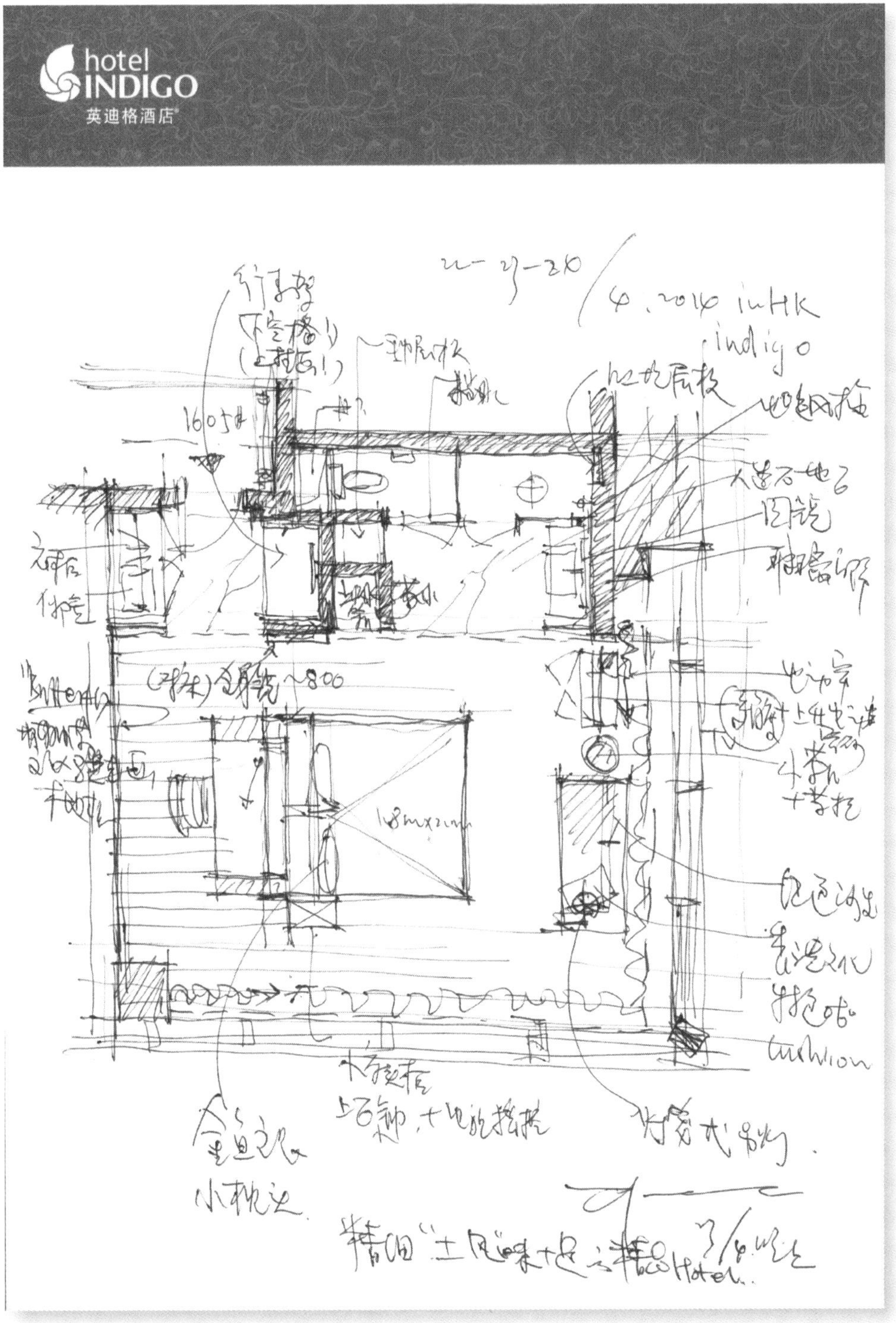
hotel INDIGO
英迪格酒店
indigo
cushion

香港港岛英迪格酒店

45 ★★★★

HOTEL INDIGO, HONGKONG ISLAND

Address : No,246 Queen's Road East Wanchai Hong Kong China (People's Republic)
香港湾仔皇后大道东246号

Telephone : +86 3926 3888

Http : //cn.ing.com/hotelindigo/hk

从诚品想到的

占据铜锣湾黄金地段奔腾了好几层的空间，诚品书店，其实这个定义不对，也不知叫它什么好：有书、家品、礼品、文具；吃的、喝的；聊天的地方、看书的地方，正如它们所叫的：诚品，就对了！

想想我们住的酒店，或者越来越不像酒店了，像"W"以家的名义来定位：大堂叫客厅，宴会厅叫起居室、书房等，也就没有了，家，酒店的定义也有所了变化，或是创意，或不得已而为之，诚品的变化不断的或是传统所致：网络书、买书的途径增加；或是互联网的冲击，成本不同，但不管怎样改变思路或经营手法都是好的，有花样总比坐而待毙强。

酒店也一样，这么多的酒店争市场，目的肯定是一大片，活下来都是有方法的，我们关注活下来的、活得好的，那才是我们学习的榜样。

向诚品的改变和坚持致敬，虽然我看到的还是入不敷出，多去看看、吃吃、买多一些书就好了；不过逛逛也行，有人气了，就有生气了！

占据铜锣湾黄金地段，希慎广场三层的空间诚品书店。其实这个定义不对了，也不知叫它什么好：有书、家品、礼物、文具、吃的、喝的；聊天的地方、看书的地方，正如它们所叫的：诚品，就行了！

Eslite Book Store is located at the 3rd floor of Hysan Square, golden section of Causeway Bay. Actually this definition is not right. I don't know how to call it. There are some places for book, home appliance, gift, stationery, something to eat, drink, somewhere to have a chat and place for reading a book, just like what the name they gave, Eslite, that is ok.

想想我们住的酒店，或者越来越不像酒店了，像“W”以家的名义来定位。大堂叫客厅、堂吧叫起居室、书房等，也颇为亲切。家、酒店的混淆也有利于营销，或是创意，或不得已而为之。诚品的多年不盈利或是传统所致，本身看书、买书的就有局限；或是互联网所致，成本不同。但不管怎样改变思路或经营手法都是必要的，有花样总比坐而待毙强。

Just think of the hotels that we had lived, they looked no longer be a hotel, like "W" which taking home standard as its target. Lobby is called as living room. Lobby bar is called as living room or lobby bar is called as study room, etc. It sounds cordial,and the mixing concept of the home and hotel is helpful to its marketing. Or maybe it is a very new idea or it has to run the way like this. For Elsite, as we all know, it has many years of non-profit as it is a traditional industry and seldom people will buy book from here or it is affected by the internet as the cost is quite different. Anyway, no matter what idea or tactic they will use, it is quite essential to change the thinking and the promoting and marketing way. Figuring out more promotional tricks is always better than just be here waiting for death.

香港港岛英迪格酒店 *Hotel Indigo, Hongkong Island*

香港港岛英迪格酒店 *Hotel Indigo, Hongkong Island*

酒店也一样。这么多的酒店争市场，死的肯定是一大片，活下来都是有方法的。我们关注活下来的，活得好的，那才是我们学习的榜样。

The hotels are the same. They also faced fierce competition. After all, the ones can survive are less. For the ones to survive, of course they have their way to survive, however, not just for survival, surviving good and making profit, which is the example we'd like to follow.

向诚品的改变和坚持致敬，虽然我看到的还是入不敷出。多去看看、吃吃，买多一些书就好了，不然去逛逛也行，有人气，就有生意了。

We hereby show our salute and respect to Eslite for their courage to change and their persistence. Although what I can see now is that they are still loosing the money, however, just have a look at or buy something to eat, buy some more books, I think that is enough for me, having a great number of consumers and popularity itself already means good business.

国际品牌长出了本土文化

在香港有些本土品牌能让我们看到一直在变的香港，以外地人包括我们经常行走的那些有印记作用，做得相当好的就有G.O.D（住好啲），意思就是住得更好！做家品的，用香港的旧照片、图案、文字、装饰做家具、布艺、包括衣服饰品，把这制作一切一切的元素有着舒服的“情结”，但这种传承一直给人以不合拍的感觉（一家之言）包括我们做设计师的，但也一直在存着！

之前入住上海的“意”得很的外滩indigo也没感觉到“海派”风，但这次香港港岛的这家酒店给人很到的“港文化”植入之感觉，好是坏，让你评价一下！

地处湾仔皇后大道东路侧的黄金地段，每层只有8间房，可想而知其“小巧”，lobby只有不到100平方米，炫的夜场变色的玻璃背景墙，二层的中空“吊挂”如彩虹的玻璃梯连接二层的中餐厅、健身房以及三层早餐间与商务区书吧，串起还很不错，有一角落作店房，体验过的特色小店所在这里走过的，几乎是酒店里所有的摆设的有特色的"港文化"的布置舍。金鱼图案、鸡公的剪纸、精美的茶具、大红漆的托盘、打太极的瓷公仔（娃娃）、旧香港照片加工的挂画、明信片、介绍图也好、怀旧年代的书本等等。这样一说，相信你们可以想象房间是怎样的了！

住在端头一间房，L形的半玻幕墙，尽览街景，"浮目胜细睇"的布局，值得学习和尝试，极尽其用，正如印证了香港设计的精髓——"以小做大"，但是设计师非常专业：硬装色调和谐，米黄大调子，色彩的小摆件将香港文化很好地浓缩于这在房间里。好不好？！哈哈，有点儿刺激！

尺度较窄，因为车位多；很嘈杂，因为楼下多；色彩多；不安宁，因为只有十几米的楼距，大大的玻璃幕墙，当然还有全开放的洗手间，好的是看起来大，但灯光、噪声的干扰还是很厉害的。或者这些都不是纪念品店。

应当承认，indigo可说是将本土文化、传统精华以轻巧地传承到我们眼里。

去探一下、体验一下，传播！

顶层的"半透"玻璃泳池也是个魔鬼般的吸引力。当然还有建筑大师Frank Grey的"扭扭"楼的华丽的顶层豪宅也给了我惊喜！

快点去，你也会置身我的感受当中的。

International Brand Mixing with the Native Culture
国际品牌长出了本土文化

在香港有些本土品牌能让我们看到一直走过来的香港，对外地人包括我们经常行走的都是有“印记”作用。做得相当好的就有G.O.D（住好啲），意思就是住得更好！做家品的，用香港的旧照片、图案、文字、数字做家具、布艺包括衣服、饰品、挂画制作，一切一切的元素有着简单的“情结”。但这种传承一直给人不高档的感觉（一家之言），包括我们做设计师的，但也一直生存着！

Some local brands in Hongkong gave its tourists or visitors the nostalgic impression of the old days. Some are doing fairy good like G.O.D (means living good). For the enterprise running the household items, use the old picture or logo or words or number as the logo or element for the furniture or clothing including the clothes, decoration, hanging picture. All these elements, simple but with sign of emotion inside. However, this inheritance has been mixing with the feeling of not-high-quality (personally speaking), designer like us is the same, however, these brands still survive in the market.

之前入住上海的“色”得很的外滩INDIGO，也没感觉多少“海”风，但这次香港港岛的这家就给人强烈的“港文化”植入之感。是好是坏，让你评价一下！

Talking about the colorful Hotel Indigo that we had lived in before, where is located in Waitan, Shanghai, I can not feel any "oversea" style. However, this hotel in Hongkong left us the strong impression of mixing with strong Hongkong culture. I don't know if it is good or not. Anyway, you can have a comment on it further.

地处湾仔皇后大道东路侧的黄金地段，每层只有8间房，可想而知甚“小巧”。Lobby只有不到100平方米，炫如夜场。变色的云石背幅墙，三层的中空，“吊挂”如彩虹的玻璃桥连接二层的中餐厅、健身房，以及三层的中餐间与商务区。书吧还很不错，有一角兼作小商店。你喜欢的特色小商品可以在这里选购，几乎是酒店里可以拿走的有特色的“港文化”的都包含：金鱼图案、艳红的靠枕，靛蓝的茶具、大红漆的托盘、打太极的瓷“公仔”（娃娃）、旧香港照片加工挂画、明信片，以及介绍周边吃、喝、玩、乐的书本等等。这样一说，相信你可以想象房间是怎样的了！

It is located in the golden section of the East Queen Avenue, there are only 8 rooms in each floor. You can guess how "smart" it is. Lobby is only less than 100 square meters, dazzling like the night scene. The discoloration of the marble back wall, hollow of the third floor, the glass bridge hanging like rainbow is connected with the dining room and keep-fit gym in the 2nd floor and dining room in the third floor as well as the commercial area. Book bar is still quite nice, and one corner is taken as small store. You can buy whatever small items there. Almost all the Hongkong characteristic culture can be taken away in this hotel, including golden fish logo, strong red pillow, blue tea set, red lacquer tray, porcelain "doll" who is practicing taiji, old Hongkong picture processing and hanging picture, post card as well as books that introducing the surrounding places like eating, drinking, playing and entertaining. From my comment, I think you can imagine what the room will be like!

住在端头一间房，L形的采光幕墙，尽览街景。“深雕细琢”的平面，值得学习和尝试，极尽其用，正好印证了香港设计的精髓——“从小做起”。倒是设计师非常老到：使硬装色调与米黄大调子和谐共存，多彩的小摆件将香港文化很好地浓缩体验在房间里，好不好？！哈哈，有点儿刺激！

Living at the end of another room, L shape lighting curtain wall is set up and all the street view can be in your eyes. The plane that is carved with exquisite carving workmanship is worth studying and having a try. It just proves the heart and soul of the Hongkong Design "Starting from trivial affair". I think the designer is quite professional, who can make the tone harmonious with the beige tone. The colorful small decoration condense the Hongkong culture very well into the room of the hotel. I am not sure if it is good or not! Haha, it is something quite stimulating!

香港港岛英迪格酒店 *Hotel Indigo, Hongkong Island*

香港港岛英迪格酒店 *Hotel Indigo, Hongkong Island*

尺度较窄，因为东西多；很紧张，因为镜子多，色彩多；不安宁，因为只有十几米的楼距，大大的玻璃幕墙，当然还有全开放的洗手间，空间是看起来大了，但灯光、噪声的干扰还是很厉害的。或者这些都不是重点考虑的。

Narrow in size as there is too much stuff. It is vey tight as too many mirrors and many colors. Never be quiet as just several meters away between the buildings, big glass curtain wall, of course, there are still the all-open washing room, which makes the space look spacious. However, the light and the noisy sound still interfere too much, or maybe all these can be neglected for the moment.

应当承认，Indigo确是将本土文化，传统精华以镜子般的方式传承到我们的脑子里。

We had to admit that Indigo did inherit the essence of the traditional culture and reflect them into our mind in terms of mirror.

表扬一下，体验一下，值得！

We need to appraise its design and always worthy of having an experience of this!

顶层的“半挂”玻璃泳池也是有魔鬼般的吸引力。当然还有它望着大师Frank Grey 的“扭扭”楼的景观，山顶豪宅也给了我惊喜！

The "semi-hanging" glass swimming pool in the top floor is still quite magically attractive. Of course, the landscape of the "shaking" building that designed by master Frank Grey, where the luxurious villa on the mountain indeed gives us infinite surprise!

快点去，你也会置身我的感受当中。

Come on, if you go there and you will share my feeling.

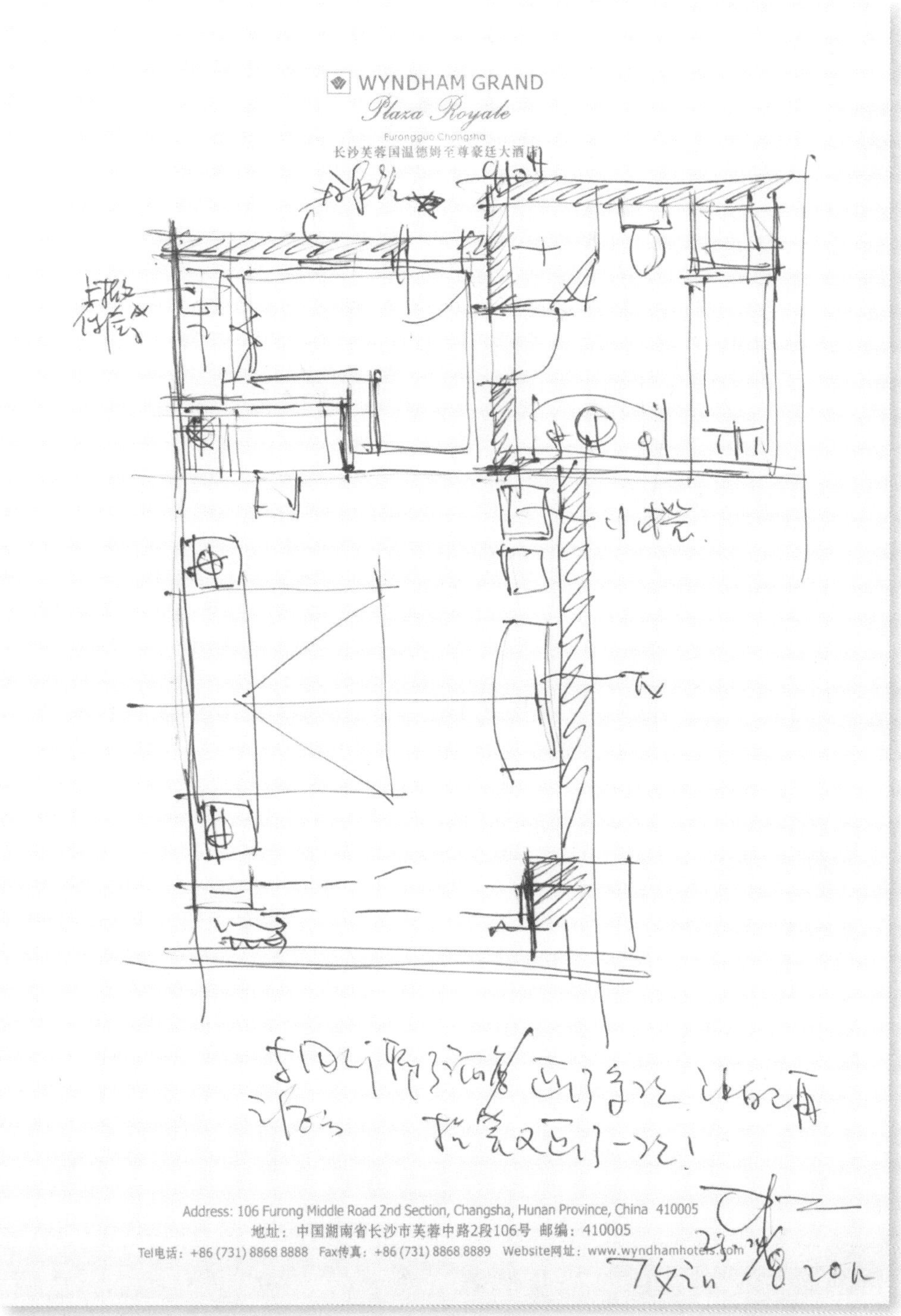
WYNDHAM GRAND
Plaza Royale
Furongguo Changsha
长沙芙蓉国温德姆至尊豪廷大酒店
Address: 106 Furong Middle Road 2nd Section, Changsha, Hunan Province, China 410005
地址：中国湖南省长沙市芙蓉中路2段106号 邮编：410005
Tel电话：+86 (731) 8868 8888 Fax传真：+86 (731) 8868 8889 Website网址：www.wyndhamhotels.com

WYNDHAM
Hotels and Resorts

长沙芙蓉国温德姆至尊豪廷大酒店

46 ★★★★★

WYNDHAM GRAND PLAZA ROYALE, CHANGSHA

Address : No.106 Furong Middle Road 2nd, Changsha,Hunan Province,410005
中国湖南省长沙市芙蓉中路2段106号 邮编：410005
Telephone : +86 731 8868 8889
Http : // www.wyndhamhotel.com/

广西沃顿国际大酒店

47 ★★★★★

GUANGXI WHARTON INTERNATIONAL HOTEL

Address : No.88,East Minzu Avenue,Nanning, Guangxi,China China (People's Republic)

中国广西南宁市民族大道88号

Telephone :+(86-771) 2111888

Fax :+(86-771) 2111999

Http : //www.whartonhotel.com/

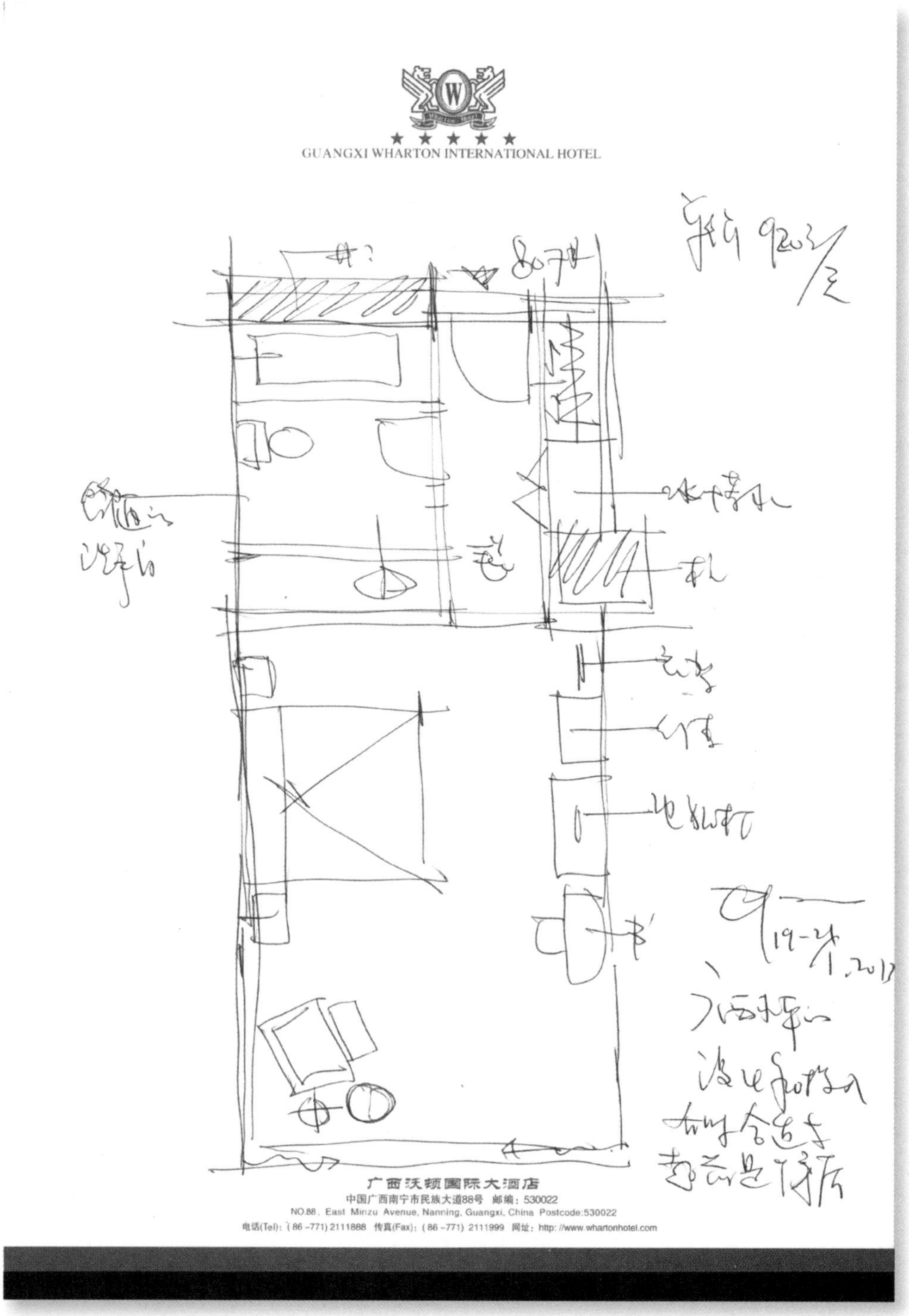

GUANGXI WHARTON INTERNATIONAL HOTEL
广西沃顿国际大酒店
中国广西南宁市民族大道88号 邮编：530022
NO.88, East Minzu Avenue, Nanning, Guangxi, China Postcode:530022
电话(Tel)：(86-771) 2111888 传真(Fax)：(86-771) 2111999 网址：http://www.whartonhotel.com

酒店的一六三

住，十个会有一个令你难忘，六十个过目即忘，三十个嗤之以鼻，你有同感吗？和我们接触项目的经历差不多，能扭转到三六一或六三一那就是大家的幸福了！

或业主要求不高，或地域定位或投资所限，设计团队水平问题，更有后期的改饰，灯光运用等原因，也会因此跟不上了整体的总评价，总之住多了，你就会更加宽容和真诚地去评价不同的品牌了。

本宁汉斯酒店从级别上或可归到最低档之列，虽然在黄金地段，但没设计，没投入，或者适合了当时当地，改造之时可以更技巧地思考差异性的投入和所带来的效益！但每个行业总有领先者，自然而然也要有对抗强手的，存在是一种勇气，也值得尊敬的，不能小看。

一六三也好，三六一也好，只要你是其中的一员，你就有成为强手的可能！

Hotel 163

酒店的一六三

住，十个会有一个令你难忘，六个过目即忘，三个嗤之以鼻，你有同感吗？和我们接触项目的经历差不多，能扭转到三六一或六三一，那就是大家的幸福了！

In terms of living, I am sure that, among 10, 1will let you never forget. 6 will be easily forgotten, 3 you promised you won't go there any more. I don't know if you share my feeling. Like the project we had experienced, if you can change it from 3,6,1 to 6,3,1, I think you are successful.

或业主要求不高，或地域定位或投资所限、设计团队水平问题，更有后期配饰、灯光运用等原因，也会因服务跟不上引来你的低评价。总之住多了，你就会更加合理和真诚地去评价不同的品牌了。

The negative comment is caused by different and complicated factors. Maybe the owner's expectation is not so high or the area positioning or investment will less fund or the poor design of the design team, the decoration in later period such as light application, etc., which will give the hotel the negative comment. Anyway, with more hotels you have lived, you will give comment to different brands in a more reasonable and sincere way.

南宁沃顿国际大酒店从效果上或可归到最低档之列。虽然在黄金地段，但没设计，没投入，或者适合了当时当地，改造之时可以更有技巧地思考差异性的投入和所带来的效果！但每个行业总有领先者，自然而然也要有衬托冠军的，存在是种勇气，也值得尊敬的，不能小看。

Guangxi Walton International Hotel can be listed as the cheapest quality in terms of effect. Although it was located in the golden road section, however, no people making the design, no investment or they need to make it suitable to the local people at that time. I think the owner should consider the individual difference and thinking discrepancy of the customers during the transformation. However, every industry should have the leader, of course, we also need the cheaper quality one to make the championed one more outstanding. Existence is a kind of courage, also it is worthy of being respected and you also can not neglect them.

一六三也好，三六一也好，只要你是其中的一员，你就有成为冠军的可能！

No matter 1,6,3 or 3,6,1, only if you are one member of them, you will be possible to get the championship!

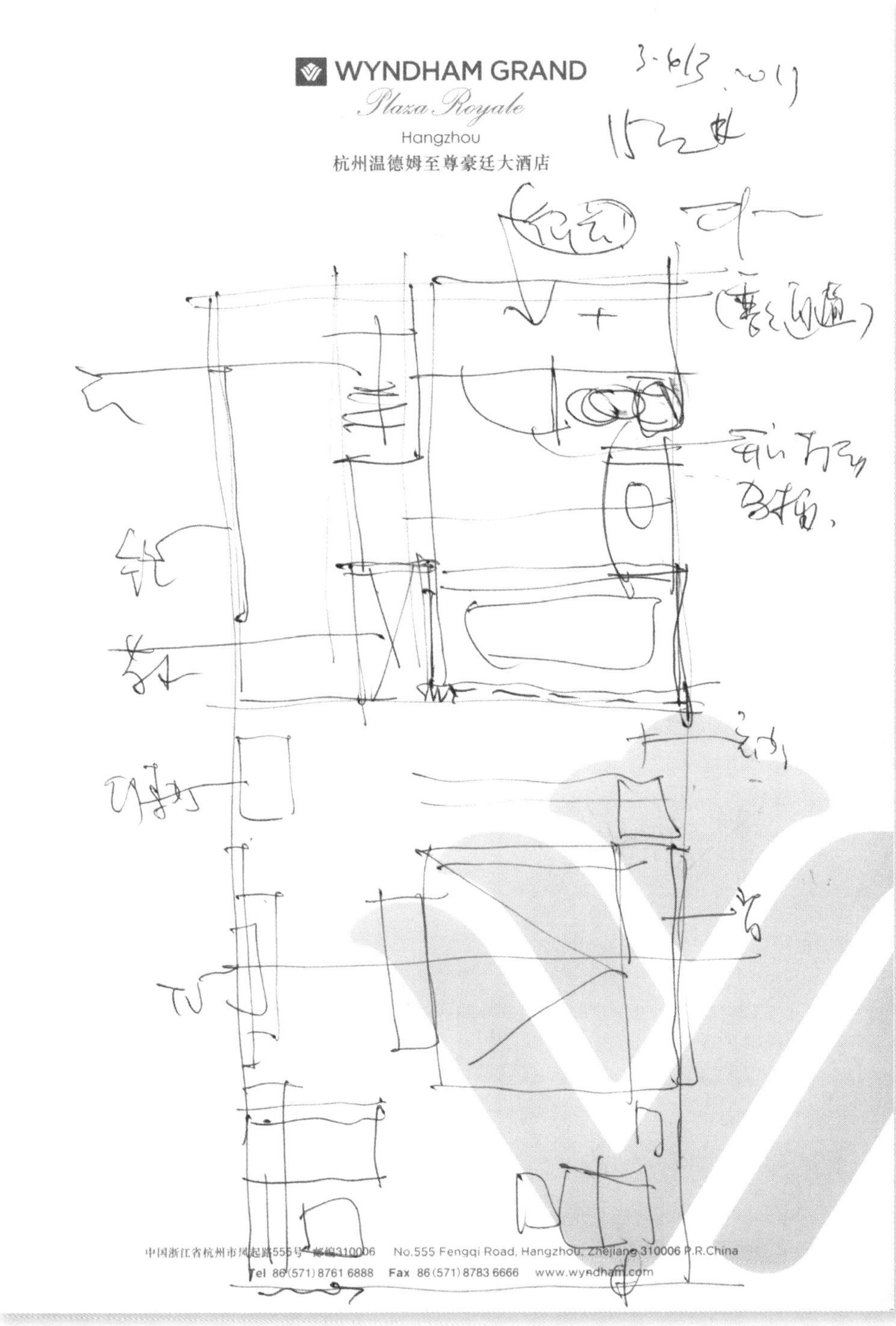
WYNDHAM GRAND
Plaza Royale
Hangzhou
杭州温德姆至尊豪廷大酒店
中国浙江省杭州市凤起路555号 邮编310006 No.555 Fengqi Road, Hangzhou, Zhejiang 310006 P.R.China
Tel 86(571)8761 6888 Fax 86(571)8783 6666 www.wyndham.com

WYNDHAM
Hotels and Resorts
杭州温德姆至尊豪廷大酒店

48 ★★★★★

WYNDHAM GRAND PLAZA ROYALE, HANGZHOU

Address : No.555 Fengqi Road,Hangzhou, Zhejiang 310006 P.R.China
中国浙江省杭州市凤起路555号
邮编 310006

Telephone : +86(571)8761 6888

Fax : +86(571)8783 6666

Http : //www.wyndham.com/

HYATT®

深圳君悦酒店

49 ★★★★★

GRAND HYATT, SHENZHEN

Address : No.1881 Baoan Nan Road,
Luohu District Shenzhen,
China, 518001
罗湖区宝安南路 1881 号
深圳，中国，518001
Telephone : +86 755 8266 1234
Fax : +86 755 8269 1234
E-mail : shenzhen.grand@hyatt.com
Http : //shenzhen.grand.hyatt.com/

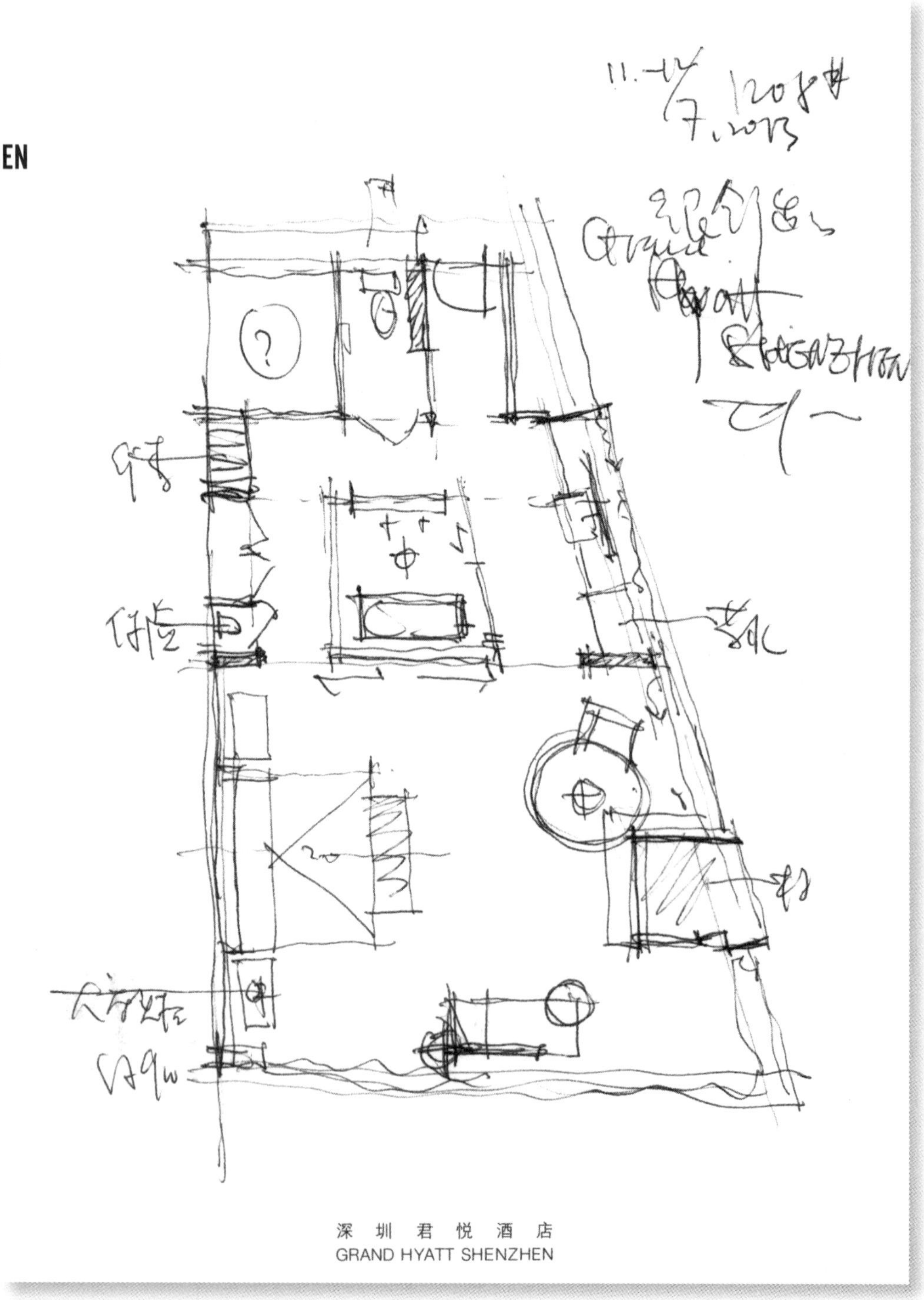

深 圳 君 悦 酒 店
GRAND HYATT SHENZHEN

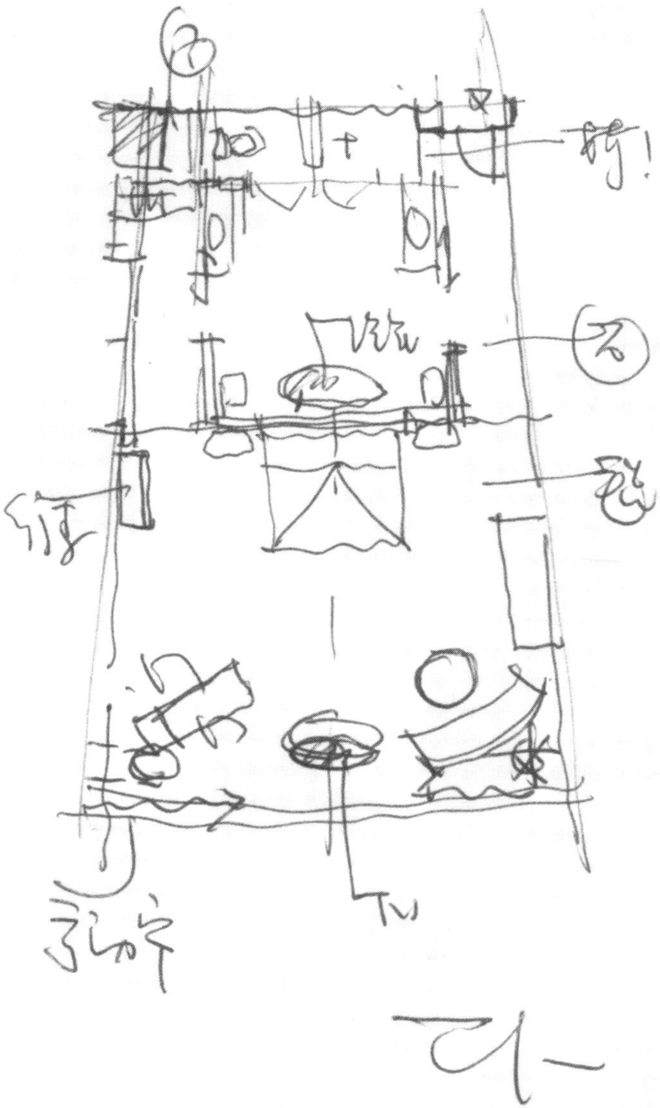

FOUR SEASONS HOTEL
Guangzhou
广州四季酒店
★★★★★

50

FOUR SEASONS HOTEL, GUANGZHOU

Address : No.5 Zhujiang West Road
Pearl River New City,
Tianhe District Guangzhou
China 510623
中国 广州
天河区珠江新城
珠江西路5号
Telephone : +86 (20) 8883-3888
Fax : 86 (20) 8883-3999
Http : //http://www.fourseasons.com/zh/guangzhou/

没住过之房型房.

广州四季酒店占据着广州现时最高之建筑物——西塔之最上端30多层，天气好的时候从70层之大堂吧俯瞰观光塔珠江之热闹，下午、晚上经常没有位置，可谓酒店一奇葩.

围绕这三角形的特殊平面，客房设计.符合中国人节节高升之心理诉求，当然更是有明显特点的结构，房间就更有趣了.每层、每间房之平面布几乎不同，包括由外围斜网状结构钢柱引起之室内空间之变化.心里想，定是花费了设计大师（HBA公司负责室内设计）折腾了一番，相信也满足了设计师之“求新求异”了.空间一流，佩服非常！

去了很多次吃饭，中餐、意大利餐、西餐、顶层的海鲜餐，也多次到大堂吧喝咖啡.公司中层活动也选这，客户的、朋友的，因为是广州的酒店，找不到什么借口去推荐给客户，借此机会研究了一下客房之平面，借用房间的便笺动之手，感觉还是不一样.还一切一的前后区分割，比例精确合理，充分利用原西头新开宽（阔）的无敌好景，无遮无挡地作休息、休闲阅读、电视、书台功能自由"散落"式地布于客厅，造型简朴，可以说第一，而色彩张力，更加合乎味道，水墨地毯更带动一阵飓风。

说到更绝，地面墙面第一种石材，少有细节及造型，可谓难得一见之佳作！

现在，酒店之酒店多了许多，这世四季可谓最好的一间，近年来。当然，华侨的大面积还是相当受欢迎之前提！

原西房的，还是相当有趣的！

A never lived Fan-shaped room
没住过的扇形房

广州四季酒店占据着广州现时最高的建筑物——西塔的最上端30多层。天气好的时候七十层的大堂吧，有点像观光塔般的热闹。下午、晚上经常没有位置，可谓酒店一奇葩。

Four Seasons Hotel in Guangzhou occupies the current tallest building in Guangzhou –More than 30 storeys on the topmost in the West Tower. If the weather turns better, the lobby bar on the 70th floor will be lively just like the sight-seeing tower. In the afternoon or at night, usually there will be no seats available and it was really quite unique and special.

圆润的近三角形标准平面向上，渐收，符合中国人节节向上的心理诉求，当然更是有利于结构。房间就更有趣了。每层每间房的平面都几乎不同，包括由外围斜网状结构钢柱引起的室内效果的变化。心里想，实实在在让大师（HBA公司负责室内设计）折腾了一番，相信也满足了设计师"求新求异"的初衷了，效果一流，佩服非常！

The rounded sub triangular standard plane extends upward and convergent gradually. It is not only compatible with the psychological demand of the Chinese people who hope to be good and better, but also more beneficial to the structure. The room is more interesting. The plane of each room in each layer is almost different including the change of the interior decoration effect that caused by the peripheral net structure of steel column. I think this idea indeed lets the master (who is responsible for interior decoration design of HBA Company) take great pains on it and believed that it has met the requirement of the master's "Pursuit for New and Change". It is really wonderful and I admired so much!

去吃过很多次饭，有中餐、意大利餐、日餐、顶层101海鲜餐；也参观了好几次客房，包括公司中层活动组织的，客户的、朋友的。因为是广州的酒店，找不到住的借口，于是推荐给客户，借此机会研究了一下"简单"的平面。借用房间的便笺动动手，感觉还是不一样，近一比一的前后区分割：明确，合理，充分利用扇形渐开阔的无敌外景，将休息、休闲阅读、电视、书台功能自由"散落"式且无遮无挡地布于窗前。造型简单，可以说单一，配色有张力，更有岭南味道，水墨地毯更引起了一阵跟风。

I had been there many times to try different food, Chinese food, Italian food, Japanese food, Top floor 101 sea food. I had visited the guestroom for several times organized by the middle level of the company or the customer's or friend's. As it was a hotel in Guangzhou, I did not find any excuse for living. Then I recommended it to my customers and took this opportunity to study its "simple" plane and layout. Use the note of the hotel room to write something feel is totally different, with one-one segmentation of the front area and the rear area : clear and reasonable, fully taking advantage of the fan-shape structure to get a gradual wider scene and eyesight. The restroom or leisure reading room, TV, book desk function is freely "scattered" before the window. Although the shape is simple, even you can say it single, the color has tension, somewhat like the taste of the Lingnan, the ink carpet even let many people to follow.

洗手间更猛，地面、墙面仅用单一一种石材，少有细节及造型，可谓难得一见的简洁！

More crazy with the washing room, it is ground wall of a single stone and it seldom had the details or shapes, quite concise and simple and never easy to be found!

现在，弧线形的酒店多了许多，广州四季酒店可谓近年里最好的一家。当然单房的大面积还是相当重要的前提！

Now many arch shape of hotel became more and more,Four Season Hotel Guangzhou was the best. In recent years, of course the single room with big area was quite important and decided the favorite of the customer.

扇形房间，还是相当有趣的！

Fan-shaped room, indeed it is fairy interesting!

east

HONG KONG

香港东隅酒店

51 ★★★★

EAST HONGKONG HOTEL

Address : No29. Taikoo Shing Road,Island East, HongKong
香港港岛东太古城道29号
Telephone : +852 3968 3968
Fax : +852 3968 3988
E-mail : info@east-hongkong.com
Http : //www.east-hongkong.com/

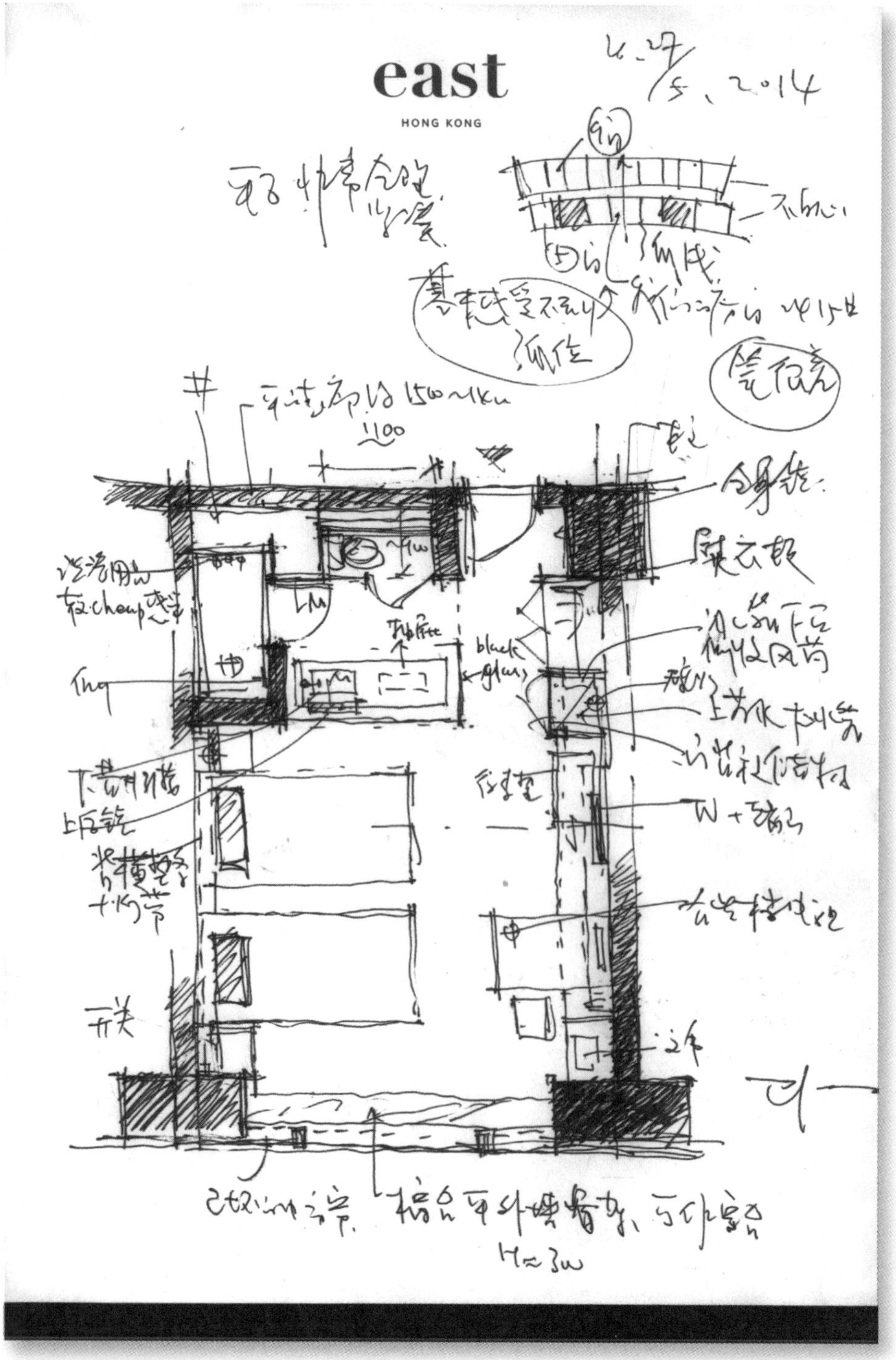

east
HONG KONG

多了

因为香港酒店设计项目的原因，匆匆忙忙地入住了位于太古城的设计型酒店——east 东隅酒店。

傲视的主楼，气派的单身而上挺的结构，尽显建筑师的功力，偏横向的客房，尺度局促，难为了室内设计师了！

因为是"公差"，和一位同事入住了双人间，此精品酒店定位可害惨我们了！

舒服，过份的设计细节，似隐似现，镜子很多，处处要小心，可能是"天书"（酒店设计标准手册）或管理公司的因素要求，房间的每一处都"尽善尽美"、"机关重重"，洗个澡生病，两个"帅哥"里里外外都不适。优点：设计细节到位，尽用空间；工艺精准、高质，不容丝毫的差错。心里想：其实这么有限的空间，有些细节，甚至某些功能性可省去（实际上使用率和舒适不高），让空间尺度更舒适一些，这样会更好！

设计不可无物，更不可多物，使用起来感觉多了。

多了或许不是设计，是居住的人。这种设计，较适合一个人或情侣般的"合一"居住。

真的，多了。

设计师的欲望！

Too Many

多了

因为香港酒店设计项目的原因，匆匆忙忙挑了位于九龙城的设计型酒店——香港东隅酒店。

Due to the project design for Hongkong Hotel, we hastily selected East HongKong Hotel, which is one Design-Characteristic Hotel that located in Kowloon Town

微孤的主楼，看似简单而工整的结构，尽显建筑师的功力。偏横向的客房，尺度尴尬，难为室内设计师了！

Slight arch of main building seems simple but in order, which reveals the power of the architecture. Horizontal form of the guestroom is quite different to make that size, which is a challenge to the designer!

因为是“公差”，和一位同事入住了双人间。时尚的精品酒店定位可害惨我们了！

As it was “an errand job”, I lived in the double room with my colleague. The positioning for becoming a fashionable elite hotel brought us the problem!

全开放、透明的洗手间区域，若隐若现，镜子很多，处处要小心。可能是“天书”（酒店设计标准手册）或管理公司的固有要求，房间的每一处都“尽善尽美”，“机关重重”。出于职业病，两个“帅哥”里里外外翻个遍。佩服：设计细节到位，尽用空间；工程精准、高质，不容丝毫的差错。心里想：其实这么有限的空间，有些细节，甚至某些功能能否省去（实际上使用率部分确实不高），让空间尺度更加舒适一些，这样会否更好?

All open and transparent washing room area is looming before you. There are many mirrors available and you need to be meticulous with care. It maybe the requirement from the Standard Manual book for Hotel Design (or we Chinese call TianShu, means the book seldom people can understand) or the inherent requirement of the management company. Every area of the room should be made “perfect and beautiful” with “details and good ideas”.Two “handsome guys” had carefully reviewed the layout and design of the hotel carefully and maybe it just because of the vocational habit. Admiration: The design is detailed to the right place, space is never wasted, the project is made quite accurate with high-quality, all of which can not be made wrong or with mistake. I can't help thinking that, since the space is quite limited, why not delete or remove some details or even some functions (actually the using rate is indeed not high) and just make the space more comfortable and better to live.

east

香港东隅酒店 *East Hongkong Hotel*

设计不可无为，更不可多为，不然使用起来就感觉多了。

Design can not be made with nothing, nor be made with too much, then you will feel better.

多了，或许不是指设计，只是指居住的人。这种设计，压根只适合一个人或情侣般的“合一”居住。

Too many does not mean that too many design elements are applied but the people who live in this hotel. That is a good thing. Well, for this design, it can only be suitable to be lived with one people or one sweet couple.

真的，多了。

Really, too many designs in return bring so many people to come.

设计师的欲望！

That's the result that the designer wants to have!

SOFITEL
LUXURY HOTELS
济南索菲特银座大饭店

52

★★★★★

SOFITEL LUXURY HOTEL, JINAN SILVER PLAZA

Address : No.66 Luoyuan Avenue
Shandong Province
250063 - JINAN
CHINA
中国山东省济南市
泺源大街66号
Telephone : +(86)531 86068888
Fax : +(86)531 86066666
E-mail : rsvr@sofiteljinan.com
Http : http://www.sofitel.com/

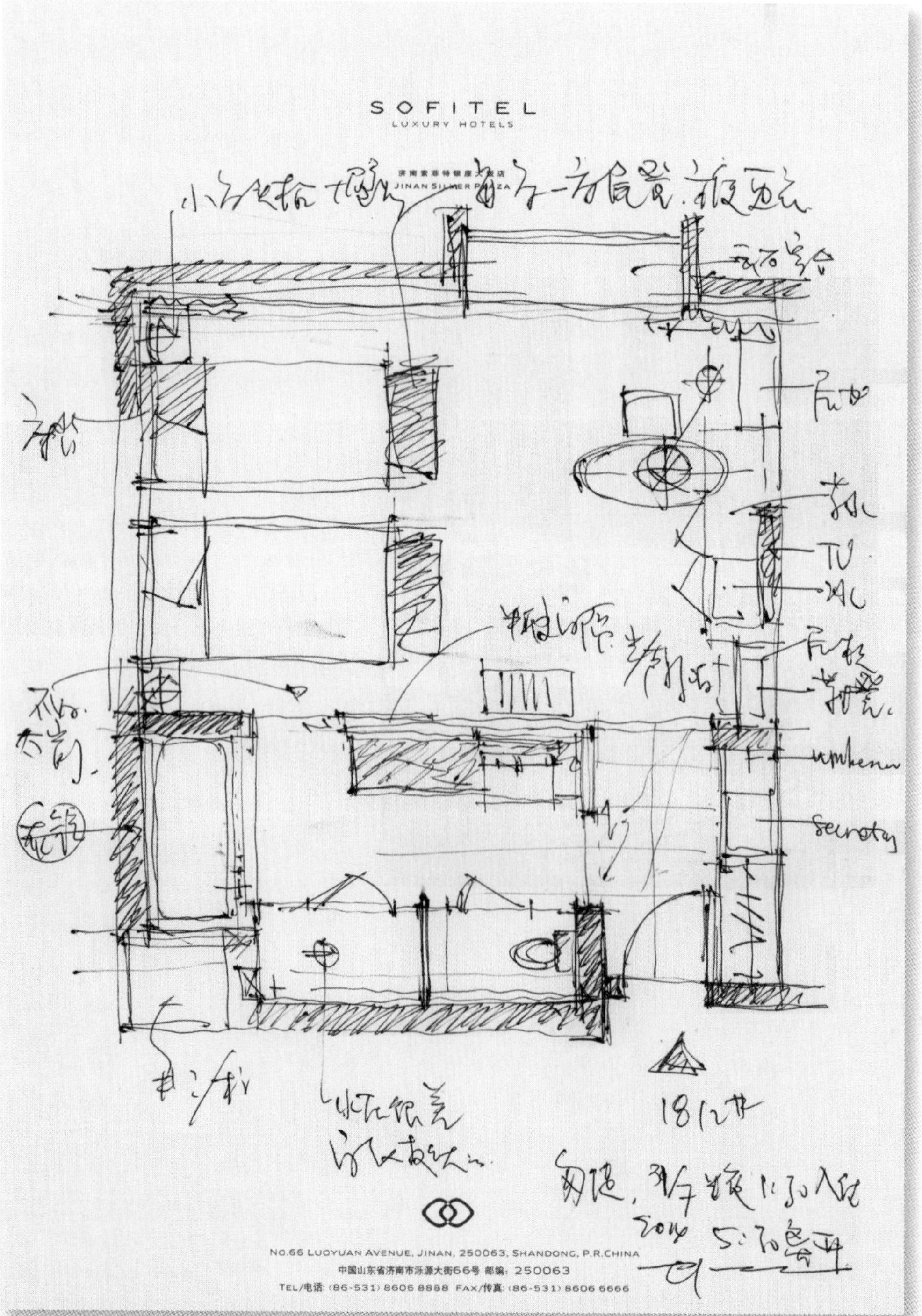

SOFITEL
LUXURY HOTELS
济南索菲特银座大饭店
JINAN SILVER PLAZA
TV
Secretary
No.66 LUOYUAN AVENUE, JINAN, 250063, SHANDONG, P.R.CHINA
中国山东省济南市泺源大街66号 邮编：250063
TEL/电话：(86-531) 8606 8888 FAX/传真：(86-531) 8606 6666

SOFITEL
LUXURY HOTELS

济南索菲特银座大饭店
JINAN SILVER PLAZA

路过　　　　　　　　　　　　　　30-31/7.晨 2:30

从本营匆匆忙忙回济南，飞半夜赶上车，抵了济南最好的王府城，最高的酒店，索菲特银座大饭店，少有会加上“大饭店”的称谓。因为要赶第二天一早下午的飞机到长沙，几个小时的住店可谓“路过”，流行的说法是“漫游”！

整体感觉还是不错，明快、时尚，可惜去毛病，房间不适合同性同住，两人，又是一个设计师忽略中方的典型案例，半透明的洗手间区，声音、光线、影子都会干扰到同住的人，相信设计师也会知道这点，但还是不遗余力地推行，有点说不过去，故当“路过”还可！

设计应当权衡了功能和经济量，“行为学”的同时也要研究心理，设计师要去住店，特别是自己设计的酒店！

小心，以其人之道，还治其人之身！

（签名）

NO.66 LUOYUAN AVENUE, JINAN, 250063, SHANDONG, P.R.CHINA
中国山东省济南市泺源大街66号 邮编：250063
TEL/电话：(86-531) 8606 8888 FAX/传真：(86-531) 8606 6666

NZONE

济南索菲特银座大饭店 *Sofitel Luxury Hotel, Jinan Silver Plaza*

Pass by
路过

从东营匆匆忙忙回济南，已是半夜一点半。挑了算是最好的、最新的五星级酒店：济南索菲特银座大饭店，少有五星级酒店会加上“大饭店”的称谓。因为要赶第二天一早7:30的飞机到长沙，几个小时的住店可谓“路过”，流行的说法是“暖床”！

When I rushed back to Jinan from Dongying, it was already half past one in midnight. I chose the best and the newest five stars hotel there, Luxury Hotel Silver Plaza, which was seldom named “grand hotel” for Five -star Hotel. Because I must fly to Changsha on 7:30 the next morning, only had a few hours staying in hotel, I could say I only had “passed by” this hotel, or had just “warmed up the bed” in a popular saying.

整体感觉还是不错的，明快、时尚，可惜老毛病，房间不适合两位同性同住。又是一个设计师忽悠甲方的典型案例：半透明的洗手间区，声音、光线、影子都会严重影响同住的人。相信设计师也会知道这个道理，但还是“不遗余力”地推行，有点说不过去，权当“路过”还可！

The whole image was nice. It's bright and fashion. But there was a common disadvantage that it was not suitable for two guests in the same gender staying together. It was a classic case which designer “tricked” consignor again. The washing room area is half transparent, so the sound, light and shadow from the one in washing room will disturb the other one outside seriously. I thought the designer should know this problem. However, they still did like that easily without excuse! It's fine for “passing by” this room!

设计应当权衡功能和欲望，“为所欲为”的同时也应当研究心理。设计师更应多住店，特别是自己设计的酒店！

Designers should find the balance between function and their desire. They designed following their desire, but also need to research the guest's psychology. Designers should stay in more hotels, especially stay in the hotels designed by themselves!

小心，以其人之道，还治其人之身！

Be careful! Maybe one day designers would be paid back for the problem they leave.

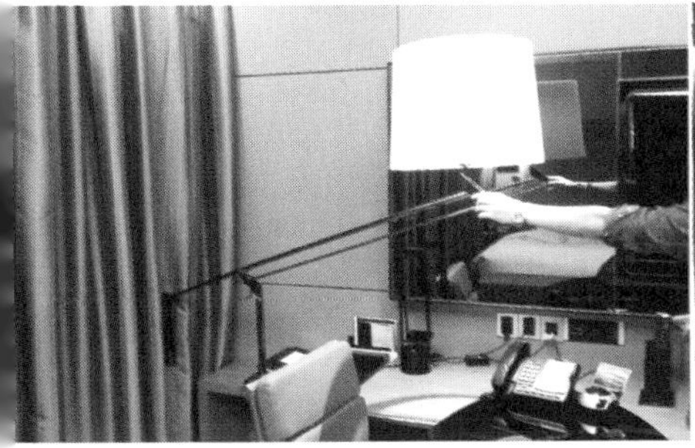

深圳回酒店

53 ★★★★

HUI HOTEL SHENZHEN

Address : No.3015,HongLi Road
Futian District Shenzhen
深圳市福田区红荔路3015号

Telephone : +86 755 88305555

Fax : +86 755 88308555

Http : //www.huihotel.com/

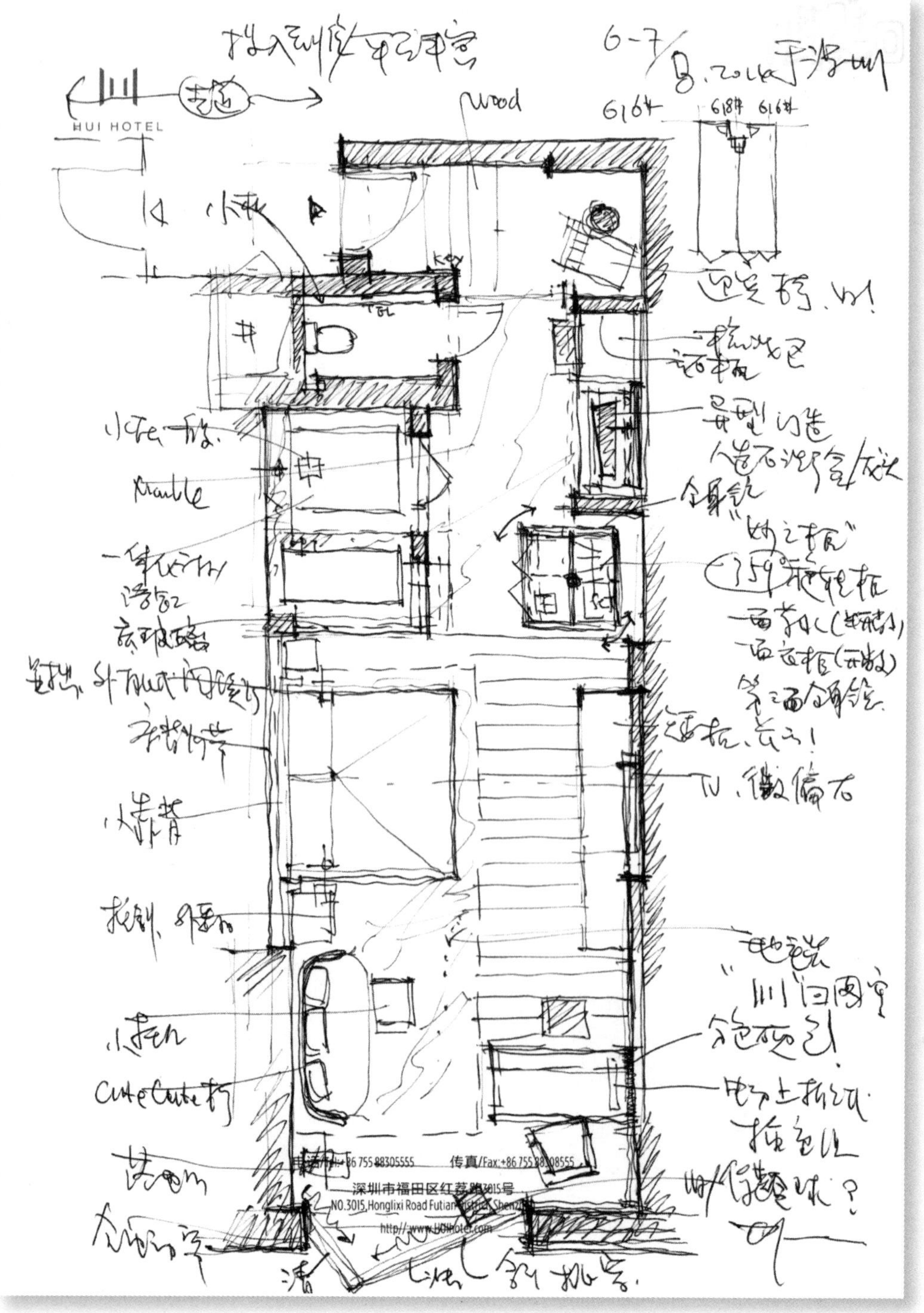

HUI HOTEL
wood
Marble
深圳市福田区红荔路3015号

设计乐一"回"

国内设计师、设计顾问参与酒店的投资故事越来越多，深圳精品酒店"HUI"回酒店，就是业内酒店设计老友黄楷郑胜先生的手脑结晶，由他更从头到尾的精心呵护，刚刚正式对外营业就引来全国同行的注目，作者也回了一"回"

闲聊中能感受到设计师对自己酒店梦想的激情，当然也心身疲累。

以其擅长（广州话叫"拿手"）的东方文化为主线，从建筑的外观改造，手段而内敛、表现；到公共空间特别是大堂和顶层加建的书吧、处处可见是设计师深思熟虑而又"随心所欲"的空间设计！

"回"到客房层的规划、布局亦是简约、大气，丝毫不逊色于顶级大品牌五星级酒店（虽然"回"酒店还没有评定星级，又或者也不会去评定），内布局是心思，

电话/Tel:+86 755 88305555 传真/Fax:+86 755 88308555
深圳市福田区红荔路3015号
NO.3015,Honglixi Road Futian District, Shenzhen
http//:www.HUIhotel.com

HUI HOTEL

对过目之客房，隐私性好，开放而灵活，令人之标准区，淋浴间使用区亦是精彩。但我相信你，会有些许争议吧。客房区的睡眠、休闲区结合外立面改造的"创意"打造的余裕面外挑窗，有效地处理了与城市马路及与其近距离对视的居民楼的关系，鼓掌功夫。我个人最喜欢走进可以随意挪动的集合了茶水服务、衣柜、全身镜于一体的设计，极富参与性与趣味性。

当然精彩的还有别致的家具和个性化的~~软装~~配饰设计等。

"回"可以说是让参与投宿的设计师狠狠地"将设计进行到底"的一回体验，"乐在其中?!"只有同行人才能感受到，不管怎样，也是设计师"梦想成真"的"一回"设计，真真正正地乐了一回?!

电话/Tel:+86 755 88305555 传真/Fax:+86 755 88308555
深圳市福田区红荔路3015号
NO.3015,Honglixi Road Futian District, Shenzhen
http//:www.HUIhotel.com

Happy Design in HUI Hotel

设计乐一“回”

国内设计师、设计公司参与酒店投资的故事越来越多。深圳精品酒店“Hui”回酒店，就是业内酒店设计的翘楚杨邦胜先生的参股项目，当然是从头到尾的精心呵护，刚刚正式对外营业就引来全国同行的注目。有幸也住了一“回”。

More and more domestic designers or design companies take part in hotel investment. The Hui Hotel, a boutique hotel in Shenzhen, is such a hotel with partly stake held by Mr. Yang Bangsheng, who is a famous hotel designer. Of course, each detail of this hotel is carefully considered by Mr. Yang, so the hotel attracted the peers of hotel all over China when it just opened. Luckily I got a chance to stay there.

HUI HOTEL

闲聊中能感受到设计师对实现酒店梦想的满足感，当然也心身疲累。

Talking about him, I can feel the fulfillment if a designer achieved their dream for hotel. Of course it's also very tired.

以其擅长（广州话叫“拿手”）的东方文化为主线，从建筑的外观改造，柔美而内敛、灵巧，到公共空间，特别是大堂和顶层加建的书吧、早餐间更是设计师深思熟虑而又“为所欲为”的空间实践！

The main theme of the hotel is oriental culture, which is designer's forte. The appearance of the building was modified to be graceful, reserved and delicate. The public area, especially for lobby, reading bar added on top floor and the breakfast canteen, was designed so considerably. Designer all made his thought come true.

“回”到客房的规划、布局亦显成熟、大气，丝毫不逊色于顶级大品牌五星级酒店（当然“回”酒店还没有评定星级，又或者也不会去评定）。布局凸显心思，入口迂回的方式，隐私性好，开放亦灵活；分合的梳妆区、洗手间使用区亦然精彩。但我相信亦会有

深圳回酒店 *Hui hotel*

许多争议吧。宽敞的睡眠、休闲区结合外立面改造时“刻意”打造的斜面外挑窗，有效地处理了与城市马路及与其近距离对视的居民楼关系，颇考功底。我个人最喜欢的是可以359°旋动的、集合了茶水服务、衣柜、全身镜于一体的设计，极富参与性与趣味性。

Back to the room layout, it's mature and great, as good as the room in top brand Five-star Hotel. Of course Hui Hotel has not got the star certification yet, or they might not join the stars certifying. The layout is considerable. Entrance is not straight so it's good for privacy, but it's also easy to open. The separated dressing area and washing area is wonderful for me, but I think it might be also disputed for others. The wide sleeping and rest area with oblique bay windows, which was specially made like that during reconstruction of the building appearance, effectually solved the problem of being too closed to city road and facing to the resident building. Such design just shows designer's skill. In personally, the one I like most is the facility with function of tea cabinet, wardrobe and full-length mirror together. It can also rotate for 359 degrees.

当然精彩的还有别致的家具和个性化的配饰设计。

Of course the chic furniture and personalized art set design is also wonderful.

“回”可以说是让参与投资的设计师狠狠地“将设计进行到底”的一回体验，“乐在其中”，只有局中人才会感受到。不管怎样，也是设计师“梦想成真”的一“回”了。

The designer who took part in Hui hotel investment "design and determine totally by himself". What an enjoyment it is! Only the one who has experience can feel it. Whatever, this time the Hui hotel made designer's "dream" come true.

设计，真真正正地乐了一“回”！

Designer is really happy here!

Postscript

后记

买到了什么？

业主出了设计费，业主买到了什么样的设计？

一种是买经验、服务、稳妥地做创新，如大多数你看到的细节，一种是买惊喜，与别不同，不同凡响，如杭州的浪漫，黑手党的Armani，所以去做也要有耐心，有一些酒店让你激动和惊喜，但很不适，一类是无惊喜但非常认同，如半岛、四季、文华东方，更有是不过不失，不痛不痒的那些也是可以接受的，如喜来登、洲际等等，这些更多印象很一般，服务跟进，装修全是成料的。

做设计也要多作准确定位，看到大公司的垄断，但其大多数是“经验之作”而非真实创意，每年才有几件值得去朝拜的大作，但投入也相当分量，总之，市场大，各种生存的方式也不一样，活出来的作品也是不同。

业主要你能买到什么样的设计了，值与不值？

What have You Bought?
买到了什么?

业主出了设计费，业主买到了什么样的设计?

The owners have paid the cost for design. What designs have they bought?

一种是买经验、服务、稳妥加微创新，如大多数你看到的酒店；一种是买惊喜，与众不同，不同凡响，如小众的安缦，黑乎乎的Amari。所以去住也应分开心态。有一类酒店让你激动和探究，如W、华尔道夫。还有一类是无惊喜但非常认同，如半岛、四季、文华东方。更有是不过不失、不痛不痒的，那当然也是可以接受的，如喜来登、洲际等等。当然更有印象很一般、服务跟不上、装修偷工减料的。

One is buying the experience and service, secure and innovative idea, just like the hotel that most of you can see. Another other is buying surprise, a feeling of special and different, just like Aman that is less popular and Black Amari. The feeling should be different when living. One kind of hotel let you be exciting and particular, like W, Waldorf. Another type of hotel gives you no surprise but very recognizable, such as the Peninsula Hotel, Four Season Hotel, Mandarin Oriental Hotel. Some ordinary one is neither good nor bad, and they are just so so. That can be still acceptable, such as Sheraton, Inter Continental, etc. Of course some are ordinary in our mind and service is also very bad and decorated with jerry material.

做设计也看看你的定位，看到大公司的垄断，但其大多数是“经验之作”。而顶尖公司，每年确有几件值得朝拜的大作，但投入也相当占分量。总之，市场大，各种生存的方式也不一样，结果出来的作品也尽显不同。

We also need to see what your positioning is when making design, see the monopoly of the big company, however, the majority of them are the customized one. Some top first-class company will offer several big works that is worth admiring and worshiping every year. However, they put into quite large amount of money. All in all, each kind of existence is totally different as the market is quite big. Of course the works are quite different as well.

业主，看你能买到什么样的设计了，值与不值?

My dear house owners, what designs will you buy? Worthy or not, anyway, it is up to you.

住住就好了!

没有像刻意去安排住哪里的酒店，继上一本《住哪里?》两年多，跑一跑可以感觉到国内的酒店业"天下大乱"，也走也乱，乱写一通，某种程度也印证了我们说自己的观点，与我的"雁过留声"。

电脑时代的影响之一是却不够重，更让人有了放弃的诱惑，毕竟这太浪费时间了。互联网、电商等已成了快捷的实践平台，我们设计行业也正受到新旧技术与观念的冲击，但又不可摆脱的落地，落之问题：人怎样电脑也要吃喝住睡，IT人也有家。是不可以在整个空间里面住吧（或者以后可以）于是就坚持到处住住，最有趣的去看看大家在做什么做出了什么!

住住就好了。

Just Live at Random!

住住就好了！

没有刻意去安排住哪里的酒店。距上一本《住哪？》两年多，理一理可以感受到国内酒店业“天下大乱”。边走边画，乱写一通，某种程度也印证了我们项目的据点与我的“雁过留声”。

We have no intention to arrange beforehand which hotel to live. It is over 2 years since my last book Where to live is published. I decided to sort out how I feel the chaos in the hotel at home. So I draw while walking, just write something. To some extent, it proves the objective of our project and my works still draw the attention of the society.

电脑时代的写写画画一点都不时髦，更让人有了放弃的诱惑。毕竟这太浪费时间了，互联网、电商等等成了快捷时尚的实践平台。我们设计行业也正受到新派技术与观念的冲击，但又不可摆脱落地、落点问题：人再怎么依赖电脑也需要吃、喝、睡、住，IT人也有家，总不可以在虚拟空间里面住吧（或许以后可以）？于是就坚持到处住住，最起码去看看大家在做什么，做出了什么！

In the time of computer and internet, write or draw something is no more a fashion. Even some people gave up doing that due to so much refreshing enticement from the internet. After all, it wastes too much time and internet and E-commerce became the practicing platform of the fashion. Our design industry is also encountered impacting from the new technology and new idea. However, we still can not avoid such an issue, that is, even computer and internet are quite popular and provides much more convenience to people, however, the problem is that, can you live in the virtual space(Maybe later can)? Then I insist on living everywhere and at least I can know what every one of you are doing and what you have done!

住住就好了！

Anyway, just live at random!

再识笔

住店，手痒痒，总想在房间里拿一点东西留念一下，大件的不行也不允许，走时研究了哪些是可以拿走的而Free的，原来也想收藏一下房间的常耗品：小肥皂，润肤露，拖鞋，洗衣袋等等，结果：小肥皂一大堆，润肤露用不及，变质；拖鞋穿一两次就起毛，不能好了；洗衣袋倒用来装东西，只有纸、笔是我的"最爱"有纪念性也实用，写写东西外，倒本还可以把玩一下，摸摸用用，比较一下档次更加高品质，更耐用的档次，也算是对酒店品牌投入的另一种品鉴吧。

或者可以评选一下酒店的笔了：田字的，文绉绉的的，华贵的，邻家近贵的，性格在的……大品牌的扎实，小众高端的有特色，有品位，遇到特别喜欢的就会多拿一支，或可以送同事，送朋友，也可满足得意人乎久的自鸣啊！

北京怡亨酒店房间里的四色铅笔是很特别，让你可以画画彩色图画；而上面的性格在酒店的金属笔杆尚沉手，有分量；好像新加坡圣淘沙的EQUARIUS酒店的也是金属的，一点也没有一次性使用的感觉，爱不释手；新开的上海文华东方的奶油色也是性感、感性，备感轻松；曼谷的玫红色、假日的嫩绿色、英迪格的鹅黄色、威斯汀的粉灰色，其因品牌的特色而设计，更可爱的是碰到几品牌的两笔几乎一样，可能是厂家的营销做太好了，又或是酒店采购的同志懒得动脑筋了，总之小小的笔让我看到酒店的细节。

同志们，藏笔应当成为住店的习惯，也让它成为真正的常规品，让酒店也感受一下我们小小的贪婪！

几年下来，你们能想象一下战利品了。

Collect a pen Again

再藏笔

住店，手痒痒，总想在房间里拿一点东西纪念一下。大件的不行，也不允许。专门研究了哪些是可以拿走的，而且是Free的。原来也想收藏一下房间的常耗品，如小肥皂、润肤露、拖鞋、洗衣袋等等。结果是：小肥皂一大堆，润肤露用不及，变质；拖鞋穿一两天就起毛，不雅了；洗衣袋偶尔用来装装东西；只有纸、笔是我的“至爱”，有纪念性，也更常用，除写写东西外，偶尔还可以把弄一下，换换用用，比较一下谁家的更加高品质，更耐用、耐候，也算是对酒店品牌投入的另一种品鉴吧。

Staying in the hotel with itchy hands, I always think of taking away some stuff to memorize that I had ever lived here when leaving. Too big size is not good. I specially study which small items can be taken away and of course something for free. I plan to collect some easy-consuming items like small soap, lotion, slippers, laundry bags, etc. As a result, piles and piles of soaps are stacked, lotion can never to be used and got metamorphic, slipper are pilling by wearing one or two days and becomes not good-looking. Laundry bag can be used for packing occasionally. Only the paper is still my "favorite", which can be used for memorizing purpose and can be used. Except for writing some words, occasionally you can still use it to compare which hotel's paper is more durable or higher quality and this can be taken as another kind of comment on the hotel and feedback to their brand investment.

或者可以评选一下排名前几位的笔了：四季的、文华东方的、半岛的、华尔道夫的、悦榕庄的……大品牌的扎实；小众的高端的、有特色、有品位。遇到特别喜欢的就会多拿一支或可以送同事、送朋友，也可谓对得起几千元的日租啊！

I'd also like to make comment on the pen from the first several hotels I had lived: Four seasons, Mandarin Oriental, Peninsula, Waldorf, Banyan Tree, etc. Big brand's paper note is always thicker. Small brand is high-end with distinction and taste. When I meet some which I like more, I will have one more or I can send one to my colleague or friend, it is worth several thousands yuan RMB of rent fee a day!

北京怡亨酒店房间配的四色铅笔很特别，让你可以画画彩色图画；上海外滩悦榕庄酒店的金属笔相当沉手，有分量；新加坡圣淘沙的EQUARIUS 酒店配的也是金属笔，一点也没有一次性使用品的感觉，爱不释手；新开的上海浦东文华东方酒店的奶油色笔也很性感，使人备感轻松；皇冠酒店的玫红色、假日酒店的嫩绿色、英迪格酒店的靛蓝色、威斯汀酒店的粉灰色配笔，实用而有各自的特色和识别性。更可爱的是碰到几个品牌的配笔几近一样。可能是供应商营销得太好了，又或是酒店采购的同志懒得动脑筋了。总之小小的笔让你看到酒店的细节。

The four-color of pencil equipped in the guestroom in Hotel Eclat Beijing is quite special,by which you can draw the colorful picture. However, the Banyan Tree Shanghai on the bund metal pen is quite substantial and heavy. by which Singapore Sentosa's Equarius Hotel also provide metal pen, which feels completely not like the disposal product and you can't help loving it. The newly open Mandarin Oriental's Shanghai cream color is quite sexy and you will feel relaxing. The Crown's rose red, Holiday's tendering green, Indigo's dark blue, Wenstin's grey pink are all practical with its own characteristic and good identification. More interesting thing is that, I found some pens that equipped by different hotels are almost the same. Maybe the supplier's marketing ability is quite good or the purchasing manager of the hotel won't make any improvement. Anyway, although pen is quite small thing, you can still feel the details of the hotel.

同志们，藏笔应当成为你们住店的习惯，也让它成为真正的常耗品，让酒店也感受一下我们小小的贪婪！

Hello, my friend, pen collecting should be made as your habit for living the hotel and let it become the real consuming commodity. Let the hotel industry feel our small greedy!

几年下来，你可以自己欣赏一下战利品了。

After some years, you can appreciate your own trophy of these small items.

记住旅程里的每一天

以前外出旅游，总是用眼看，用相机拍，时常记得很少。相片回来又看了几遍？？相片像纸一样，日子到了每次旅程的记忆也就淡忘了，当然精彩的部分是不会忘却的，特别是当你一起随行的人，和一起经历的事。

某次旅程中突然，除了画画写生，干脆每天晚上把当天剩下的精力留给当天的记录。于是就有了"旅游日记"，这样下来也积累了一小堆了。

这样，酒店的信纸用上了，画画，当然是用房间里的纸和手边带来的笔；酒店的便笺用上了，而且每天日记，刚好！有时候心情激动澎湃，洋洋洒洒，一本叠便笺很快用完，用半幅纸的篇幅来写要求多一些，继续记录好一天的见闻，突然也难忘了自己了。

里面提到的人，或再有联系，或已经陌生，但再次阅读画面如历眼前；里面提到的事，或觉有趣，或觉幼稚，回想旅途，或觉不可思议，心情又比当年的轻松；里面当然也记录了自己的"不晓人事"的肚子"

一样的海鸥事件，回想更是心生感觉，我去过就是为记住你们，让你们记住每一天，让你的回忆更完美，有字有图，让你的旅程更添乐趣。相信我，这个会好玩的！

不信？！你可以试试，让记忆印在纸上，而不仅是在相片里，眼睛里，脑袋里……

美化生活，那才快乐。

Remember Every Day in Your Journey

记住旅程里的每一天

以前外出旅行，只是用眼看看，用相机拍拍，脑袋记住多少，相片回来又看了几遍？相信你我一样，日子长了，每一次旅程的记忆也就淡忘了。当然精彩的部分是不会忘却的，特别是与你一起牵手的人和一起经历的事。

Going outside for Journey before, we just saw by eyes and recorded it by camera. How much we can remember in our mind and how many times we reviewed the pictures we had taken after return? I believed, with time passing by, every journey became not clear. Of course the wonderful part you will never forgot, especially the one who took your hand to experience the life and something you had experienced together.

某次旅程中想想，除了画画平面图，干脆每天休息前用当天“剩下”的体力写写当次的事，于是就有了“旅游日记”。这样下来也积累了一小堆了。

Some idea came into my mind, except drawing the plane & layout design picture, why not use my "remained energy" to write the incident that happened at that time, that is what you can see now "the Tourist Diary". Accumulating day by day, now I can keep quite a lot of diaries.

这样，酒店的信纸用上了，画画图，当然是用房间里面顺手牵来的笔；酒店的便笺用上了，可以写写日记，刚好！有时候心情澎湃，洋洋洒洒，一大叠便笺很快用完。用“半桶水”的英文来要求多一些，继续记录好一天的见闻，想想也难为自己了。

Thus, the note of the hotel can be used finally. I draw a picture on it, of course I use the pen in the room of the hotel. The note of the hotel can be used now to write the diary. At that time, my heart was so exciting. I wrote and drew something on the note and stacks of notes are running out very soon. Use my "poor English" to write more and continue recording one day's chores. Speaking of persistence, I think it is a hard job and seem too much for me.

里面提到的人，或再有联系，或已经陌生，但再次阅读，画面如历眼前。里面提到的事，或觉滑稽幼稚。因是旅途，也记录了多次的“不省人事”的“吐子”一样的酒醉事件，回想更是心里发笑。或者这就是日记的作用，让你记住每一天，让你的回忆更充实，有字为据，为你的旅程平添乐趣。相信我，这个会好玩的！

For the people it has mentioned or keeps in touch once again or it has already been strange, but if you can read it once again, the picture is still showing before your eyes. For the case it has mentioned inside, some feels funny and childish. As it recorded the things in the journey, also it recorded so many affairs that I "got drunk" and "vomited", I think it was still funny when looking back, or maybe this is the function of writing diary and it will let you remember each day and let your memory more fulfilled as it has words for evidence. It adds me interest to my journey. Just believe me, this will be very funny!

不信？！你可以试试，让记忆印在纸上，而不单单是在照片里、眼睛里、脑袋里……

Believe it or not? You can try, try to print your memory on the paper, but not just in the picture or in the eyes or just in the mind…

等你老了，那才是宝贝了。

Especially when you grow older, that will be the most precious items.

Tour Diary

旅行日记

旅游日记

美国八天

注：选取美国八天游的日记，整理，原汁原味还原当时的心情和事件；删除部分比较敏感或不宜的片段。

8 Days' American Trip

美国八天

注：选取美国八天游的日记，整理，原汁原味地还原当时的心情和事件。删除部分为比较敏感或不宜的片段。

Note: This is the diary selected from 8 days, American Trip. Now it is under sorting out. I will restore you my true feeling at that time and some events happened at that time and of course delete some sensitive or inappropriate section.

第一天，从上海飞芝加哥。

匆匆和客户在浦东机场喝了杯果汁，基本谈好了(福建)福鼎和另一个城市的地下管廊的初步合作协议，算是完成了一件事！

约了一点钟集合，原来是4:10分的机，临时还通知迟到[illegible]多小时起飞，傻乎乎在机场耗了近四小时，倒是碰到了一位当红的男性韩星引起小轰动的事情！

一年内第二次去美国，这次去东部，谢谢栩勤，给了这次机会！

十二多个小时到达芝加哥，可惜好印象一下子没有了，在没有信号的关门前苦苦等了三个小时，或者说波士顿爆炸事件所累，机场仍然下着半旗，愿逝者安息！

半小时的晚饭，回到酒店——帕尔玛酒店(Palmer house)希尔顿旗下系列，一人一房，相当好，单位不错，独立私人空间，但没有那些人性化而自然的安排。

飞机上邻座的是一位上海威尔逊公司的主设计师，还好，工作十七、十八年还有激情，Jane，从设计师转到设计管理，是李总越来越像个设计人在管理及其他领域中悟到了快乐，悟到了比设计更重要的东西。有待进一步了解才能证实！

澡，简单整理一下行李，准备入睡，洗漱用品齐全的。忘了上次美国旅行的洗澡了，这可是一间有名的古酒店啊，可惜太累了，不然可以到处看看！

同志们，早安了！

零点，20—12/4，2013

美国时间。

1、2/ 帕尔玛酒店的大堂设在二楼，可谓真正意义上的高端、大气、上档次，大工业时代的繁华与技术表露无遗、金碧辉煌。历史悠久而风采依然。

3/ 典型的“大美”装修风格,硬朗而实用。

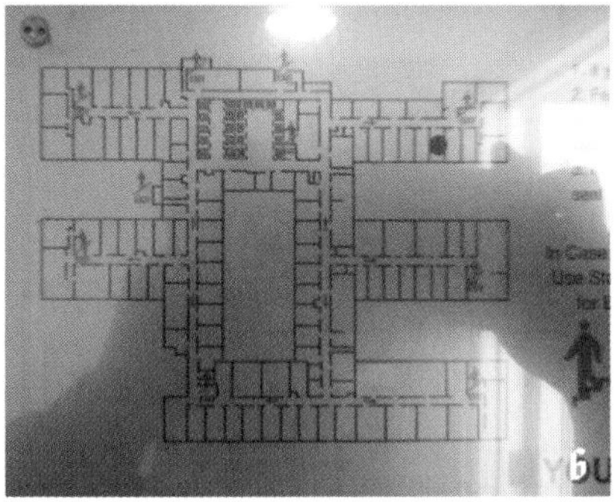

4 / 无缝不锈钢“大豌豆”成了芝加哥的象征，也落俗一下，留个烈日下的“艳照”。

5 / 团友们聚精会神地听导游的讲解，力求在最短的时间里“听懂”这座城市！

6 / “大酒店”的气派，“迷宫”一样的客房层。

7 / 外国人的幽默，将从我们紫禁城“挖来”的屋檐宝贝镶嵌在繁华大道的建筑物的墙上，让我们看得不是滋味！

The First Day. flying from Shanghai to Chicago

第一天，从上海飞芝加哥

匆匆和客户在浦东机场喝了杯果汁，基本谈好了（福建）福鼎和另一个城市的地下商场的初步商务协议，算是完成了一件事！

约了一点钟集合，原来是4:10的航班，结果还通知推迟半个多小时起飞。傻乎乎在机场耗了近四小时，倒是碰到了一位当红的男性韩星引起小轰动的事情！

一年内第二次去美国。这次去东部，谢谢科勒，给了这次机会！

十二多个小时到达芝加哥，可惜好印象一下子没有了。在没有信号的关闸苦苦等了三个小时，或者拜波士顿爆炸事件所累，机场依然下着半旗，愿逝者安息！

半小时的晚饭，回到酒店——帕尔玛酒店（Palmer House Hotel），希尔顿旗下系列，一人一房，相当好，单价不高，独立私人空间，组织方作了相当人性化而自然的安排。

飞机上邻座的是一位上海威尔逊公司的女设计师，还好，工作十七八年还有激情。Jane，从设计转到设计管理岗位。是否越来越多的设计人在管理及其过程中悟到快乐，悟到了比设计更重要的东西，有待进一步了解才能证实！

洗个澡，简单整理一下行李，准备入梦。洗澡用品超全，忘了上次美国旅游的状况了。这可是一间有名的老酒店啊！可惜太累了，不然可以到处看看！

同志们，早安了！

00:20—2013.4.19美国时间

8、9 /　大师Frank Grey (弗兰克・盖里)的大作，露天的音乐广场和“延伸”的跨公路大桥的高投入提升了城市形象，可谓钱用在最必要的地方！

10~12 /　年轻的而领先的工业城市里，高楼林立，古典而富吸引力，随手一拍就是很好的“学习资料”。

8

9

I drank a cup of juice hurriedly in Pudong Airport with customers, and we basically talk about the preliminary business agreement on the underground shopping mall in (Fujian Province) Fuding City and another city, having finished one important affair.

Assembly on about one o'clock, I realize that 4:10 is the time for on-boarding,however, later I was still informed delaying taking off for half an hour.I stayed at the airport for nearly 4 hours and didn't know what to do. However, one famous Korean star suddenly appeared at the airport and caused a small sensation!

10

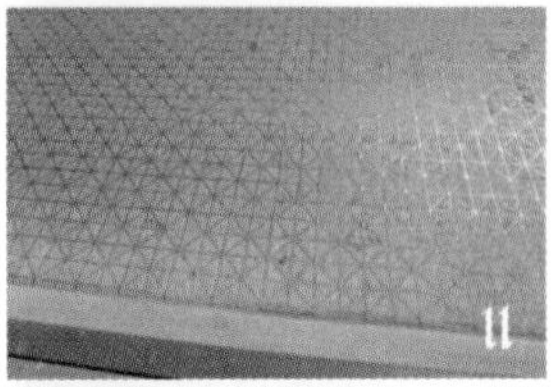

11

12

It was my second time going to U.S. in one year. This time I went to the east, thanks Kohler for giving me this chance!

It took me more than 12 hours to arrive in Chicago. What a pity that good impression was disappeared so soon. I had to wait for 3 hours at the gate without any signal. Or it was caused by Bombing in Boston, however, the airport was still under the flag and may the dead rest!

Half an hour's dinner, then back to hotel-Palmer House hotel, which is the hotel subordinated to Hilton Hotel, one room for one people. It is fairy good and price is not expensive. It is equipped with private room. The organizing party did it fairy humane and natural arrangement.

My neighboring seat is one female designer from Shanghai Wilson Company. She worked for seventeen or eighteen years, however, she is full of passion. Jane, who was transferred from design to design management. I wondered if more and more designers can feel the happiness during the management or the process of the management or something more important than the design itself. This should be conducted further investigation and verified before drawing a conclusion!

Take a bath and simply sort out my luggage, then fall into sleep. My bathing items are full and I almost forgot the condition when my last time toured the U.S. This is one famous old hotel! What a pity on me is too tired, otherwise I would like to have a look at elsewhere and everywhere!

Hello, my friend, Good morning!

00:20. April 19th, 2013 U.S. Time

第二天。

倒时差的一夜醒了很多次，早上（当地时间）想去游泳，后来担心体力不支而放弃了。七点半去与大堂同层的餐厅吃早餐，原来很多的团友已经吃完，倒时差确是奇妙！甚的认识了我们多几个团友，一边交流，轻松享受到了美国的新的一天的开始！

今天芝加哥一天游，相当的深刻的记忆，特别是弯曲的豆子（我自己这个别名更形象），无缝的不锈钢，太神奇了。精彩。弗兰克·盖里（Frank Grey）的露天音乐厅，大桥，一个比一个刺激我们的神经！下午的行程是平淡，但内河游船观光含英文的讲解也可津津听之，也是好事！晚游在密歇根湖，看到不一样的城市！

紧接着飞到第二个落脚地新奥尔良，很晚很晚的晚餐，大堂酒吧之出品，但相当丰盛是进一步认识了更多的同行人，同行的设计师，其中上海HBA公司的Monica，从事软装的台湾美女；北京的Hassell公司的华裔美女，中国地区的发家人；米勒的高档品牌Kallista的杨震先生；相互增进了解，也深入地交流。

对了，差点忘了中午的美味大餐了，是一间有相当历史的牛排餐厅，传统的大餐：沙拉，主菜，甜品，非常新鲜的原材料，14oz（安士）的大大的肉，可吃撑了，开始了"大食之旅"了！

谈到世界级公司的管理，世界级境外公司在中国的生存状态，人才培养与后续，虽然与之比较我们也有优有劣，但存在的问题共通性还是相当一致。

加油，中国的同行！

1 / 芝加哥著名的“双玉米楼”（Marina City）玛丽娜城，号称世界上最豪华、最奢侈的停车场。配有公寓、购物中心、办公、影院及游艇码头。

2 / 近看这个可爱的东西，步移景异，还真的是“简单而有趣”！

3 / 酒店的早餐不错，居然有中式的油条，看起来走遍全世界也能吃到至爱的食品了。

WWW.RENTCOWLIMO.COM

4 / 自由国度的表现，穿街过巷叫卖自己的创意！当然是卖广告的啦！

5 / 爱吃肉的我，将最有名的牛排奉上，鲜嫩多汁。

The Second Day
第二天

倒时差的一夜，醒了好多次。早上（当地时间）想去游泳，后来因担心体力不支而放弃了。七点半去与大堂同层的餐厅吃早餐，原来很多的团员已经吃完。倒时差确是奇妙！其间多认识了我们几个团友，一起交流，轻松享受到了美国新的一天的开始！

今天是芝加哥一天游，记忆相当深刻。特别是弯曲的“豆子”（我自己这个形容更形象），无缝的不锈钢，太神奇了，精彩。弗兰克·盖里（Frank Grey）的露天音乐厅、大桥，一个比一个刺激我们的神经！下午的行程则显得平淡，但内河游船观光全英文的讲解也可练练听力，也是好事！畅游在密歇根湖，看到不一样的城市！

紧接着飞到第二个落脚地新奥尔良。很晚很晚的晚餐，大堂酒吧的出品。但相当高兴的是进一步认识了更多的同行人，同行的设计师。其中上海HBA公司的Monica,从事软装的台湾美女，以及北京的Hassell公司的华玮美女，还有科勒的高档品牌中国地区的当家人Kallista杨露先生，相互增进了解，也渐渐融洽。

对了，差点忘了中午的美味大餐：沙拉、主菜、甜品，非常鲜嫩的原材料，140oz（安士）的大大的肉，可吃撑了，开始“大食之旅”了！

饭间聊到公司的管理，也聊到境外公司在中国的生存状态，人材培养与后续。虽然与之比较， 我们也有优有劣，但存在的问题的共通性还是相当的一致。

加油， 中国的同行！

6、7 / 入住的新奥尔良希尔顿酒店，相对简朴，像我们北方的建筑一样，采暖的出风口在靠窗位置。

8 / 飞抵新奥尔良，入夜城市的璀璨灯光与优美的机翼如此相衬。

It's a night for jet lag, so I woke up many times during sleeping. In the morning of local time, I wanted to go swimming, but then I gave up because I worried about being too tired. At half past seven I had breakfast in canteen which was on the same floor with the lobby. At that time many tourist members already finished eating. It's amazing for jet lag! During breakfast I got acquainted some more tourist members and talked to each other. We all felt a relaxing new beginning of a day in America!

Today we had a tour around Chicago. It gave me some impressing memory. Especially for the camber bean, I described it myself like that, the stainless steel of seamless was amazing and wonderful. Then we visited the open-air pavilion designed by Frank Grey and the great bridge, which excited us more and more. The journey in afternoon was common, but it's also a good thing that I could practice my English when listening the guide during the boat trip. Travelling in Michigan Lake, I saw a different city!

Then we flied to our second destination New Orleans. We had dinner very late, which was from lobby bar. I still felt happy because I got familiar with my tourist partners, who were designers, too, including Monica from HBA Shanghai, a Taiwan lady who was soft-mounted designer, Ms. Hua Wei from Hassell Beijing, and Mr. Yang Lu, who was president of Kohler China and Kallista, the high brand of Kohler. We knew each other better, and became more harmonious.

Oh, yes, I almost forgot mentioning the delicious lunch. Salad, main courses and desert were all made of very fresh material. The meat on 140oz made me so full. What a "big stomach" trip I had!

KALLISTA
ANN SACKS

During meal we talked about company management, the business status of foreign companies in China, and staff training and holding. Comparing different companies, we had our own advantage and disadvantage, but the problem we had was similar. I believed we could work hard and reduce the distance between the foreign companies and us.

Work hark, our designer peer in China!

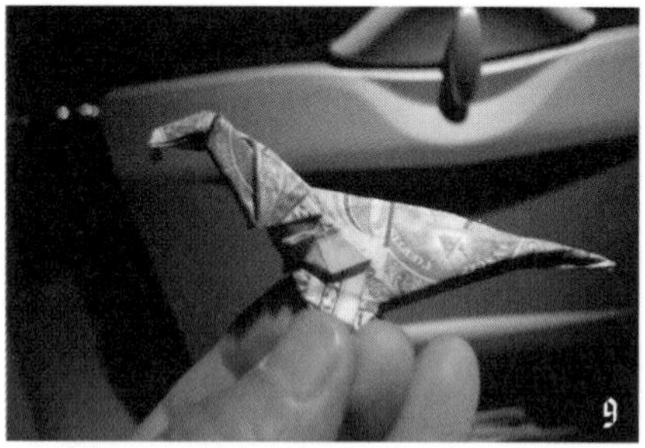

9 / 飞机上，按照杂志的指引，折了一只小恐龙，“价值”一美元，不含手工费。

10~12 / 厨卫展（KBIS），科勒展位成为焦点，其中的带智能播放的浴缸以及高档品牌，名师设计的KALLISTA都吸引眼球。当然由“肥妞”示范的有滑动挡板的浴缸最得我喜欢！

13／ 新奥尔良市中心街头艺术家的大作，个性尽显！

14~16／ 必不可少的景点游玩，人气相当旺。

从芝加哥飞新奥良

从芝加哥飞新奥良，入住很晚，吃盒饭，近十一点。

有一位女士几乎在机场内不见，走失了，所以学英语很重要，但我学了几年还是这个水平，有时和别人较真真不容易！

早上，在希尔顿酒店开始。

下午起来就去跑步了，因为酒店在机场旁边，所以定位不高，后面就是"农村"——脏乱差，很多条铁路，包括在用的和已废用的，所以导游说这个城市也叫"废都"，不为过。

早餐和博那维公司的刘女士聊天，人不错，像"管家婆"一样有才。这天上午去看卫浴展，KBIS，这次外出美国的借口，伊州本土的老大，展览展品、通过接待、解说都是"了得"，佩服。其中大师级的高档系列，如芭芭拉·芭莉的系列，又如能播放音乐的淋浴房系列。印象最深的是由一位胖嘟嘟的美女示范的浮动挡板的浴缸，方便肥大人士和老人家使用淋浴，确确实实高出对手许多，应当成为经典。

午饭在市区的热闹地，皇家街（Royal Street），精心找了一家人气最好的转角位置的餐厅，街边临窗的美丽午饭，几十美元三人吃海鲜，有鱼有虾，非常超值。两台，六个团友，轻轻松松，聊天，晒大太阳，慢慢吃，非常轻松的一天！

很晚的飞机，飞到了和我们三亚很像的地方，度假天堂：迈阿密。

在房间打开电视就看到房地产项目的广告，七百万美元一间公寓（相当于我们的高层住宅），价格吓人，存了一下，回国可以和我们的客户交流交流。

迈阿密希尔顿酒店

1 / 著名的餐厅里的招牌美食，有虾、有鱼。

2 / 三人一台的休闲，阳光大餐。你看，美女帅哥笑得多开心。

3 / 早上跑步，酒店后面的“农村”，残旧的铁路。

VIZCAYA
VIZCAYA

4~6 /　意大利文艺复兴风格的著名豪宅——维斯卡亚花园，是很多美国电影的取景地点。我们也当了一回明星。

Flew from Chicago to New Orleans

从芝加哥飞新奥尔良

从芝加哥飞新奥尔良，入住很晚，吃完盒饭已近十一点。

有一位女士几乎没有在机场内见到，走失了，所以学英语很重要。但我学了几年还只是这个水平，有时和自己较真真的不容易！

早上，在希尔顿酒店开始。

7:30起床就去跑步了，因为酒店在机场旁边，所以定位不高，后面就是“农村”——脏、乱、差。很多条铁路，包括在用的和已废用的。所以导游说这个城市也叫“废都”，不为过。

早餐和杨邦胜公司的刘女士聊天。她人不错，像“管家婆”一样，而且有才。这天上午去看卫浴展，KBIS，这次有外出美国的借口，作为本土的老大，展览、展品、组织、接待、解说都“无对手”，佩服。其中大师级的高档系列，如芭芭拉·芭莉系列，又如能播放音乐的淋浴房系列。印象最深的是由一位胖嘟嘟的美女示范的有滑动挡板的浴缸，方便肥大人士和老人家使用沐浴，确确实实高出对手许多，应当成为骄傲。

午饭在市区的热闹地——皇家街（Royal Street）用餐。精心找了一家人气最好的转角位置的餐厅，可欣赏街边临窗的美丽景观，几十美元的三人台消费，有鱼有虾，相当超值。两台，六个团友，慢慢地、轻轻松松地聊天、晒大太阳，享受相当轻松的一天！

乘很晚的飞机，飞到了如我们三亚般热的地方，度假天堂：迈阿密。

在房间打开电视就看到房地产项目的广告，七百万起的一间公寓（相当于我们的高层住宅），简约、时尚，拍了一下，回国可以和我们的客户交流交流。

迈阿密希尔顿酒店
2013.4.21-22美国时间

I flew from Chicago to New Orleans, and check in at the hotel very late. When I finished lunch, it's nearly eleven o'clock.

One lady missed in the airport. She should be missing her way. So learning English is quite important. However, I have learnt several years and still in this level. Sometimes I found it was never easy to be strict to myself.

In the morning, I started to live in Hilton Hotel.

At 7:30, I got up and went out for a walk. The hotel is not so high as the airport is just nearby. At the back is the "countryside- all is dirty and messy. Many railways including the in-use or not-in-use are lying there. Therefore, the tourist guide made a joke that this city is a "Demolished Metropolitan".

I had a chat with Ms. liu From Yang Bang Sheng Company during the time of having breakfast. She is nice, just like a "housekeeper" and she is quite talented. We went to visit the Kitchen & Bathroom International Show (KBIS). This is just an excuse for the American. As the native people, she is quite outstanding in terms of exhibition from displaying product, organizing, receiving or interpreting. We are so admired of her. Among of them is the high-quality product range such as Babara Berry Series. It is the shower room that not only can play the music. The deepest impression on my mind is that one fat beauty who make demonstration in the bathtub for showing the convenience of the fat people and old people to take a bath. Indeed it is quite competitive than its rival and should be proud of that.

We had Lunch in the busy downtown, Royal Street, carefully select one popular restaurant at the corner, which has beautiful landscape near the window, three people desk and afford about several dollars for consumption only. There are fish and shrimp available, it is really worthy of that. Two people desk and six tourists, we can sit there chatting relaxedly, sleep in the sun, slowly and slowly, It is quite relaxing day.

The late plane takes us to the sweltering place like Sanya, which is a holiday paradise: Miami.

Just open the TV and you will see the advertisement of the real estate project, seven million for one apartment (it is about the same price as house price in China), simple and fashionable. We took picture there so that we could have a further communication with our customers at home.

Hilton Hotel, Miami

April 21th-22th, 2013American Time

7 / 迈阿密海湾巡游后，到酒吧试着自己制作的Mojjto（墨吉托）鸡尾酒，好紧张的样子啊！

8 / 游轮上的兄弟姐妹，灿烂的阳光，晒得大家都睁不开眼。

第四天

还是在迈阿密，早餐很丰富，和深圳的赵伟俊男士同桌，原来之前在美院（广州美术学院）的交流会上已见过。随后去参观有名的维斯卡亚花园（Vizcaya Museum Garden）可惜室内不准拍摄，但实在漂亮，只能偷偷摸摸地。100多年的建筑，颇为濒海而建，美好至极！

中午是中式自助餐，还好！小憩一会

午后是游轮观光，沿水道看到很多很多有钱人名人的别墅，还是被震撼到了。虽然我不喜欢移民，但中国人要出争气，更要改变一下与人相处之道，让外国人更加尊重我们。要像以前那样，努力工作，努力赚钱，改变生活环境，不要只是抱怨。

晚上继续海鲜大餐，鱼虾蟹应有尽有，吃撑了！倒是之后去看Setai和W酒店感到有收获的，细节、灯光、环境都很棒，不枉此行！

同去参观的一共七人，在Setai酒店消费了一下，相互也聊得挺多的。看来开放的心才有开放的态度，有更开放的交流，谢谢同行的人！

1

2

The Fourth Day

第四天

还是在迈阿密，早餐很丰富。和深圳的靓仔纹身男王冠同桌，原来之前在美院（广州美术学院）的交流会上已见过。餐后去参观有名的维斯卡亚花园（Vizcaya Museum Garden）可惜室内不准拍摄，但实在漂亮，只能偷偷摸摸地拍，100多年的建筑，濒海而建，美妙至极！

中午是中式自助餐，还好！小憩一会。

午后是游轮观光，沿水道看到很多有钱人、名人的别墅，还是挺受刺激的。虽然我不喜欢移民，但中国人应当争气，更应改变一下与人相处之道，让外国人更加尊重我们。设计师更应是这样，努力工作，努力赚钱，改变生活环境，不应只是抱怨。

晚上继续海鲜大餐，鱼、虾、蟹应有尽有。吃撑了！倒是之后去看Setai酒店和W酒店挺有收获的，细节、灯光、环境都很棒，不枉此行！

同去参观的一共有七个人，在Setai酒店消费了一下，相互也聊得挺多的。看来有开放的心才有开放的态度，有更开放的交流，谢谢同行的人！

3

4

1~4 / 晚上慕名去参观迈阿密Setai酒店、W酒店，当然要消费一下，享受一下五星级的服务。

I was still in Miami today. The breakfast was substantial. I sat in the same table with Wang Guan, who had tattoo and from Shenzhen. Before that I had met him in an exchange meeting of Guangzhou Academy of Fine Arts. After breakfast we visited the famous Vizcaya Museum Garden, it was really beautiful but not allow to photo, so I could only take photo slinkingly. Such a building with over 100 years' history was constructed by sea. How wonderful it was!

The lunch was Chinese buffet. Just all right for a break.

We had boat trip in afternoon. I saw many villas of rich man or famous person, and got some shock from that. I don't like immigration but I think Chinese should win credit for themselves, and change the way of getting along with others, so that the foreigners would respect us more. Even the designers should do like that. They should work hard, earn money hard, and improve the life condition, but shouldn't complain all the time.

The dinner was seafood again. There were fishes, shrimps and crabs. I was so full! After that, visiting Setai Hotel and W Hotel made me feel a lot. All details, lights and environment are great. It's a nice trip!

There were totally seven persons who joined visiting, and consumed in Setai Hotel. We talked a lot each other. Having an open heart and open attitude could bring more communication. Thanks, my tourist partners!

辛苦的一天

辛苦的一天，但大家都很开心，那就是血拼（shopping）。

早上美味的水果早餐后，9:30开车一个小时来到郊外号称北美最大的Outlet店，索格拉斯折扣店中心（Sawgrass Mills）确实气派，导游事先介绍了一下购物攻略，体量、环境、货品都相当吸引，大小姐们"短暂"旅程开始了，大家早就都做足了功课，拟好了购物清单，除了我，所以我也要找点乐子才行，很快在第四区找到了美甲的店，太棒了！逛了一条街，买了一个耳机，是降噪性能的，试了一下，不好用，会掉出来，再买了一个硬的耳机，这个好一些，"大功告成"。

中午和几个团友吃大餐，绿树、[illegible]、大虾、意粉、香肠、蘑菇沙律，很美味。之后也认真逛了一下，再花时一个半小时做了手指、脚指的保养按摩，和国内的享受差不多。

满足的[illegible]之后开车去游湖河（船上观光）很放松的旅程。

回到酒店已近十一点，明天一早4:30起床，一天下来真累！ 大家晚安！

2/6，2013

5

NAIL TRIX
Thank You!
6

5 / 拍我们的美女，互拍。
迈阿密的沙滩，有椰子相伴，好开心啊！

6 / 折扣店里面的修甲空间，让我打发了时间，也享受了异国的服务。

Working Hard for A day

辛苦了一天

辛苦了一天，但大家都很开心，那就是血拼（shopping）。

早上美味的水果早餐后。9:30开车一个小时来到郊外一号栋北美最大的Outlet店：索格拉斯折扣中心（Sawgrass Mills），确实气派。导游专门介绍了一下购物攻略，体量、环境、货品都相当吸引人。六小时的“短暂”旅程开始了，大家原来都做足了功课，有购物清单，除了我，所以我也要找点乐子才行。很快在第四区找到了美甲的店，太棒了！逛了一个多小时，买了一个耳机，是弹性胶的，试了一下，不好用，会掉出来。又买了一个硬胶的，这个好一些，“大为充实”。

中午“抓”了几个团友吃大餐。绿树阳光下，大牛扒、意粉、香肠、蘑菇沙拉，很美味。之后也认认真真地逛了一下，再耗时一个半小时做了手指、脚趾的保养按摩，和国内的差不多。

满足的同行，之后开车去游船河（船上观光）。这是一个很放松的旅程。

回到酒店已近十一点，明天一早4:30起床，一天下来真累！

大家晚安！

2013.4.23

7、8 / 阳光下的牛扒大餐，当然谢谢美女、帅哥的赏脸，户外的架空花园空间，好有情调的哦！

Working hard for a day and all people were happy. That was shopping.

After enjoying the delicious fruit during the breakfast in the morning, we set out at 9:30. It took me one hour to arrive in our biggest outlet store in North America located in the suburb Lot 1, Sawgrass Discount Center (Sawgrass Mills), indeed splendid. The tourist guide specially introduced the shopping strategy, measurement, environment, goods, which were all quite attractive. Six hours' "short" trip begun, obviously all people had prepared enough homework for this. All people had the shopping list, except me. I think I need to find some fun now. I had found it very soon in manicure's shop in Block fourth, it was really great! I had strolled there for more than one hour. I bought one headphone, which is one elastic rubber. I tried it on but it still didn't work. The elastic rubber always drops. Then I bought another one with hard rubber. It's bigger and better.

In the afternoon, I "caught" some teammates to have dinner together. Under the green tree and sunlight, big steak, spaghetti, sausage, mushroom salad, all were tasted delicious. After that, I strolled carefully and it took me one hour and a half to have a finger and toe massage. It was almost the same as at home.

Our tourist partner seemed quite satisfied, then they drove car to cruise in the river (sightseeing on the boat). It was one relaxing journey.

When came back to the hotel, it was nearly 11 o'clock. Early 4:30 tomorrow morning, I'll get up. Indeed it is a very tired day!

Good Night!

April 23th, 2013

又回到了芝加哥

又回到了芝加哥，飞机落地，顺利，顺路去看了科勒的展厅，这个展厅位于一幢小建筑内，都是建材的，非常高档，或者是它的产品就是这样的，包括科勒的展厅在内，每家的面积都相当大，科勒的更加有品位，特别是展架，展示形式和展示效果，非常值得中国的建材企业的借鉴，包括我们，特别是自然、精细而优雅的，非同一般！

中午吃芝加哥特色的批萨，老店，气氛、装饰、出品都很好，当然包括我们的热情，真是充满笑声的盛宴，给午餐丰盛的！

餐后，互相讨论，没有继续去血拼（shopping）去了一个小型的Mall，再一次展现了我们巨大的购买力，一车一车的战利品，又一次添加了很多的新买的箱子，相当便宜，148美元，两个，结果超级壮观，大客车的行李箱和部分座位都是箱子！

之后参观了JW万豪酒店，对面的W Hotel也顺便进去转了一小圈，万豪酒店经典而又植入了新的现代化的改造，也是"古典翻新"的一种好方法。W酒店很潮，不宜久留，而万豪人气十足。这就是区别：太个性，只能小众；大家都认同才能众乐乐，这个道理或者适合所有的行业！

吃午中餐之后去看show，叫"Blue Man"，非常出名的演出，也很美国式的幽默，当然观众也是关键的互动，我们的美女华华就是其中的参与者、表演者。这个经历，相信对她是值得纪念的。尾段更是高潮，全场舞蹈和"打球"活动（大大的气球）

太累了，11:30开始休息，懒了，所以这是第二天早上才写的！

[illegible] 早上，2013

1~3 / 芝加哥的W酒店，在万豪对面，"顺路"去看了看。改造的酒店，艺术当道，时尚张扬。

CONNIE'S
PIZZA
1963
CHICAGO
STYLE PIZZA

又回到了芝加哥，飞机顺利落地。顺路去看了科勒的展厅，这个展厅位于一栋旧建筑内，都是卖建材的，非常高档。或者这里的品质就是这样的，包括科勒的展厅在内。每家的面积都相当大，科勒的展厅更加有品位，特别是展架，展示方式和展示效果非同一般，非常值得中国的类似企业借鉴，包括我们，特别是为追求精细而付出的人们。

中午吃芝加哥特色的披萨。这是一家老店，气氛、装饰、出品都很好，当然也包括对我们的热情。真是充满笑声的盛宴：简单而丰盛！

餐后，我们互相讨论，改变计划继续去血拼（shopping）。去了一个小型的Mall，再一次展现了我们强大的购买力，一车一车的战利品，又一次添加了很多秀丽的新箱子，相当便宜，148美元两个。结果颇为壮观，大客车的后备箱和部分座位都是箱子。

之后参观了JW万豪酒店，对面的W 酒店也顺便进去看了一小圈。万豪酒店经典而又植入了新的现代化设备，也是“古典新造”的一种好方法。W酒店很暗，不宜久留，而万豪酒店人气十足，这就是区别。太个性，只能独乐乐，而大家认同才能众乐乐，这个道理或许适合所有的行业！

享用了简单的中餐之后去看show，叫“Blue Man”，是非常出名的演出，也很有美国式的幽默。当然观众也是关键的互动，我们的美女华华就是其中的参与者。这个经历，相信对她来说是值得纪念的。尾段更是高潮，是全场舞蹈和“打球”（大大的气球）活动。

太累了，11:30开始休息，懒了。所以这是第二天早上才写的！

2013.4.25早上

Returned to Chicago

又回到了芝加哥

PUGET

4~6 / 又折回芝加哥，到了一家正宗的披萨店去享用午餐，开心得很。多种多样的配料。看看大家的笑容就可想而知有多么好吃了。

Returned to the Chicago and the plane landed, all went smoothly. I was on my way to Kohler's Exhibition. This Exhibition is located in one old building. They sold the building material, very high-quality. Or the quality here is just like that, including showroom of Kohler. Each showroom is quite big area, Kohler has more taste especially the display rack, way of display and display effect, which all are worth taking as reference for the Chinese counterpart , including us, especially for the people who has pursuit of fineness and perfectness, just unusual and fantastic!

In the afternoon we had the Chicago characteristic Pizza. It is an old store in atmosphere, good decoration and taste, of course including our passion, banquet filled of laughter: Simple but rich and abundant!

After dinner, we discussed with each other and changed as going shopping, then went to one mini-Mall. Once again displayed our strong purchasing power, a car of booty, once again added many new beautiful boxes, quite cheap, two pieces only USD148. You can imagine the most extravagant scene, the trunk of big passenger car and part of the seat were fulfilled with boxes.

After that, we visited the JW Marriott Hotel, and W Hotel is just in the opposite. So we dropped in as well. JW Marriott Hotel is classic and implanted with new and modernized equipment. Also it is a good way of "classic design". W Hotel is quite dark, not good for staying long, however, Marriott is quite popular, Which is the difference. If it is too individual, we can only take it for fun. Only the highly recognition make people happy. This is a simple reason and criteria, which I think it is suitable for all the industries!

After finishing simple lunch and then we went for a show called "Blue Man". It is one famous show and it is an American way of humor. Of course it is the key that the audience can be better responsive to it. Our beauty Buahua is among one of the participants. This experience, I think it is one memorable experience for her. The tail section is the climax, all the dancing performance and "playing ball" activity (big balloon).

I am too tired. Time comes to 11:30, so tired, so it is written in the morning of the next day!

JW MARRIOTT
7

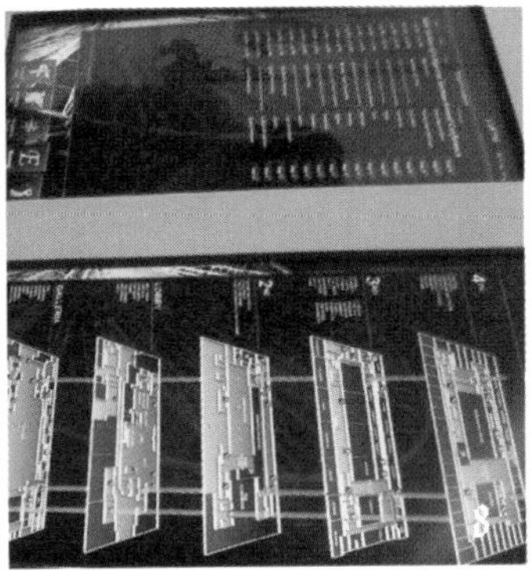
8

In the Morning, April 25, 2013

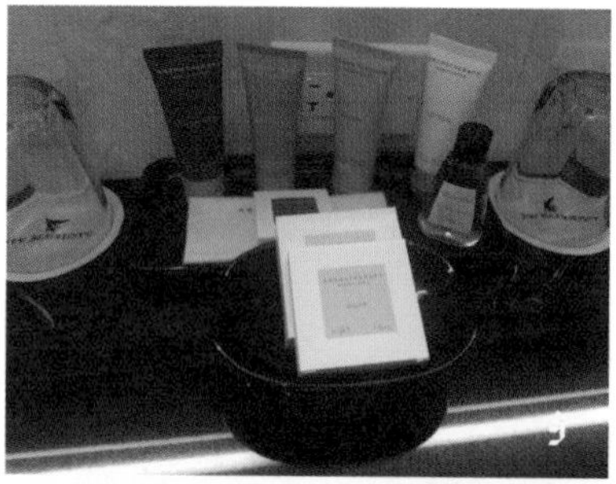
9

10

11

12

7~10 / 万豪酒店有相当的历史，暖色为主，与W酒店形成较大的反差，但与时俱进的科技引入，也使其保持市场的竞争力。

11 / 著名的表演，BLUE MAN。演出后观众积极与主要演员合影留念。

12、13 / 芝加哥科勒主展厅位于一幢绿色认证的建筑内，连卖场的公共区域都这么雅致，可想而知入驻的商家的素质和定位。

VISIT
13

14

THE BOLD LOOK OF KOHLER
KALLISTA
robern
KOHLER SURFACES
15

16

STAIRS
17

18

14、15 / 科勒展厅综合了熟悉的洁具（多个品牌）、面板、瓷砖，琳琅满目，黑钢铁的笔挺陈列架，我最喜欢，值得学习，当然要有很好的工艺和加工。

16 / 这样整洁的陈设方式，你服了没？反正，我服了。

17 / 酒店里，一群刚刚游完泳的美女帅哥在自娱自乐（帕尔玛酒店）。

18 / 另一间的帕尔玛希尔顿客房，果绿色的装饰，特别提神。

科勒公司140周年

因为昨天参加140 Years Kohler（科勒公司140周年）酒会，总之所以说起这次酒会，喝多了，因此今天早上补写的！

结果和澳洲一样成了"吐记"，还是没有冲到洗手间就吐了，地毯又遭殃了！这次进步了，还写了一张纸条在床头连同小费，告诉他们可以从信用卡里扣钱，如果有需要的话！

早上好好地舒服地享受了Kohler的产品，先是玻璃洗脸盆，冲水式马桶的淋浴间（竟产品的人想出来的，一点用都没有，水力分配不均）不过淋浴的门构造超棒，完全防水外溢设计特好！然后是好好地泡了一下按摩浴缸，看了十分钟的画片，好像是《[illegible]》，有点说笑的。

早餐也有统一的时间，因为大家都很晚，是下午式的早餐：一个鸡蛋，两片培根，咖啡一杯，一盒水果，一杯橙汁。

昨天的140 years Kohler酒会让我见识了140岁的企业之风采和气质，特别是全同行尊重的理由："挑战、坚持、沉淀、创新"！

回顾，昨天从芝加哥早餐后就来科勒镇，离芝加哥只有两个多小时，很美的一个地方。中午去"黑狼奔"高尔夫会所吃饭，餐厅很朴素：石、木，之后去参观了小地方：马圈、呼啸峡球会，接着入住科勒的酒店——the American Club Resort，之前是他们的员工宿舍，晚餐之时去逛了一下附近的商场，顺便买了一双美国手工的皮鞋，以尊重晚上的宴会（酒会）

其实科勒某种程度上也和中国大多数企业一样，有钱了就圈地、盖房子、搞艺术，做公益，目的只有一个——更好地生产利润，良性循环。这就是企业的责任；对社会的责任！

20/6.2013. [illegible]

1/ "黑狼奔"高尔夫会所午餐，景色比食品更精彩。

2~ 4 / 科勒的酒店，旧员工宿舍改造，惹人喜欢。当然也是科勒集团的接待和推销产品服务以企业文化的最好载体。

床头柜上的小松树很特别，也许是对绿化的渴求，全屋的Baker（贝克）家具（也是科勒集团属下的品牌），经典得不得了，当然是原汁原味的美式设计，舒适得很。

Kohler Company's 140 Anniversary

科勒公司140周年

因为昨天参加140 years Kohler (科勒公司140周年)酒会及之后的泡吧庆功会，喝多了，因此日记是早上补写的！

结果和澳洲一样成了“吐子”，还是没有冲到洗手间就吐了，地毯又遭殃了！这次进步了，还写了一张纸条在床头。连同小费，告诉他们如果有需要的话，可以从信用卡里扣钱。

早上好好地、舒服地享受了Kohler的产品。首先是玻璃洗脸盆，冲水式多样的沐浴间（卖产品的人想出来的，一点用都没有，水力分配不够）。不过沐浴的门构造超棒，高端而防水外溢的效果特好！然后是好好地泡了一下按摩浴缸，看了十分钟动画片，好像是《里约热内卢》，挺满足的。早餐也有充足的时间，因为大家都睡得很晚，所以通常是下单式的早餐：一个鸡蛋、两片培根、一杯咖啡、一盆水果、一杯橙汁。

昨天的140 years Kohler酒会让我见识了140岁的企业的风采，确实有令人、特别是令同行尊重的理由：“持续、坚持、沉淀、创新”！

回顾，昨天从芝加哥早餐后就来到科勒镇，离芝加哥只有两个多小时的车程，是一个很美的地方。中午去“黑狼奔”高尔夫会所吃饭，餐厅很朴素：石、木。之后去参加了马圈、呼啸球会。接着入住了科勒的酒店——the American Club Resort,之前是作为他们的员工宿舍。空余之时去逛了一下附近的商场，顺便买了一双美国手工的皮鞋，以表示对晚上的宴会（酒会）的尊重。

其实科勒某种程度上也和中国大多数企业一样，有钱了就圈地、置房产、搞艺术、做公益，目的只有一个——更好地产生利润，良性循环。这就是企业的责任，对社会的责任！

2013.4.26早上补写

创·新创
Bold.Art.

140
KOHLER
Bold. Design.

5 / 为表示对主办方的晚宴的尊重，买了一双本地的手工皮鞋，经典款式，喜欢它的舒适、粗犷。

6~8 / 有幸出席科勒的140周年庆活动，看到令人尊敬的老科勒先生。简洁而震撼的现场，令人忘怀，当然热情的招待也十分令人满意。

I write this dairy in this morning as I drank too much yesterday when joining in Kohler Company's 140 Anniversary Cocktail Reception Party as well as the celebration party in the pub.

I drank too much and kept vomiting like an Australian rabbit(the pronunciation of rabbit is the same as word "vomiting" in Chinese). The carpet is ruined! This time, I became smart enough. I wrote some words on one paper slip together with the tips and put them on my bed in order to tell them that the money can be deducted from my Credit Card if necessary.

I can enjoy Kohler product comfortably in the morning. First of all, it is glass facial basin, shower room with varieties of water flushing method (the people who sell this did not consider this problem, it is no use at all and the distribution of water force is not strong enough), however, the door of the showroom is super nice in terms of construction and it is high-quality and has good water-proof and overflowing-proof effect. Then I just sleep on the massage tub and saw ten minutes of animation cartoon somewhat like "Rio de Janeiro", I felt quite satisfied. Also there was enough time to have my breakfast. As we worked very late, usually the breakfast is ordered in advance: One chicken egg, two pieces of Bacons, a cup of coffee, a bowl of fruit, a cup of orange juice.

Yesterday's 140 years Kohler cocktail party let me have the opportunity to meet this elegant enterprise of 140 years old, especially the only reason for their peer to show respect, which is the business tenet they have been keeping "sustainability, persistence, precipitation, innovation"!

Just recall things happened yesterday, I finished breakfast and then came to Kohler Town. It is only 2 hours away from Chicago and it is one beautiful place. In the afternoon I went for lunch at "Black Wolf Run" Golf Club. The canteen is quite simple, just stone and wood. Then I joined the horse ring, Whistling Straits Ball Club, then lived in Kohler Hotel-the American Club Resort. Before it was their staff's dormitory. I went around the shopping mall nearby during my spare time, and then bought a pair of U.S. handmade shoes for tonight's banquet (cocktail).

To some extent, actually Kohler is like most of the Chinese enterprises. When the company has the money, they will buy the land, build the real estate, make arts and do charity. The only purpose is how to produce the profit in a better way, thus make virtuous cycle. That is the responsibility of the enterprise, also their social responsibility!

It is written in the morning of April 26th, 2013

ARTS CENTER

9 / 科勒艺术中心原是科勒先生的住家，经过改造后成一家对外开放的小型公众建筑物，也算是为家作的贡献，外国人的做事方式真有很多值得我们去学习的。

10 / 下午，艳阳背景下的贝克（baker）家具展示，各个阶段的经典椅子，一一陈列，受益匪浅。

11 / 艺术中心展品不少，我最喜欢的其中一件，全部是用纸做的，张扬而神秘。

12 / 去小木屋的路上随手拍，春末的蓝天与未发芽的树干、常绿的树交错成美妙的图画。

13 / 老餐厅、老陈设、老的羊皮书。

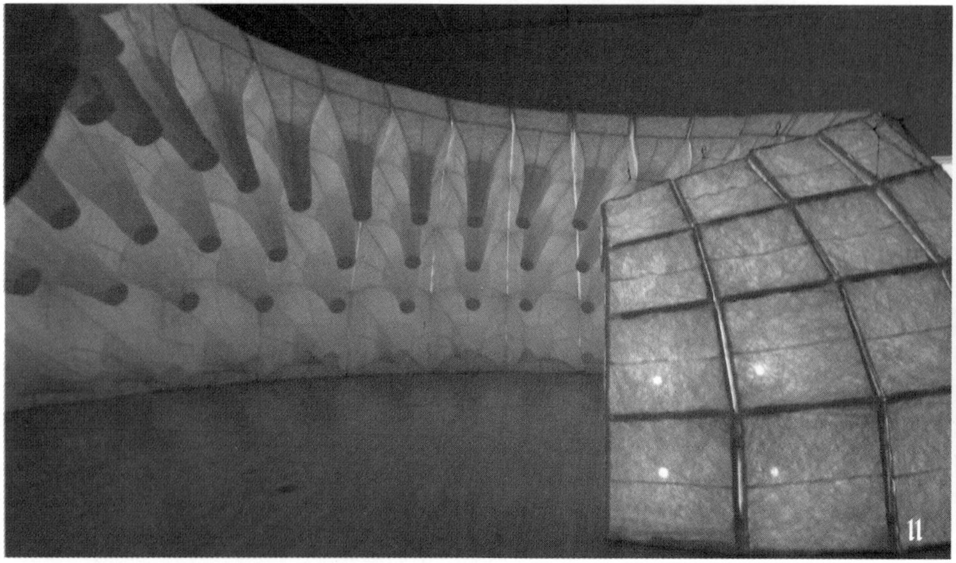

最后一天

算是最后一天在美国的旅行，因为明天一早就去机场，10点多飞机回上海，早上的科勒镇参观了设计中心，之后去到科勒艺术中心，展品和展馆非常精彩，特别是用了很多、很多、很多"瓷"做成的装置，相信艺术家要耗时近一年才能完成，对我的设计周展览有不一样的启发，"相信艺术，相信生活，相信市场·移动·移动"这或者可以作为以后艺术定位的口号。

向科勒学习，将艺术引入"生活"，这样可以提升品位和印象。

中午在科勒的木屋吃饭，环境一流，出品不合口味，自认为很差。

四个多小时车程回到芝加哥的帕尔马酒店，住了第二次。自由活动的时间，选择还是看看名店和酒店。

著名的密歇根大道，名牌店，七点已渐渐关门，只能外观！著名酒店永远是开门的，沿途都是经典的，看了很多家：Westin，Drake（Hilton的古典品牌），很古典，精致

酒店

最后选择了在柏悦吃饭，很欣赏它的经典设计，大开眼界。再约到同团的伙伴，一齐吃饭了。很好，因为周末，餐厅没位，只能坐酒廊，也很如愿，正如帅哥说的，是这次美国之行"最像样"的一餐。那就是，品牌就是保障（正如科勒一样），我们也希望做到这样。

餐后慕名找芝加哥唯一间米其林三星的餐厅：Alinea，很难找，就是一间私人屋，很难看，不知道出品会怎样。相信只是将食艺术化，吃后说，开始不相信米其林的标准了，这样打了分，相信就是玩技巧，过了些才吸引眼球。就像设计一样，相信很快就会不行的，我相信的是如科勒一样的"老金"，沉淀最重要！

4/10.2017

BEST BUY
Chase

1 / 半下沉式的广场，都是吃的、喝的。这间餐厅的门面相当有吸引力。

2 / 到处都有“苹果”。

The Last Day

最后一天

这算是最后一天在美国的旅行，因为明天一早就去机场，乘10点多的飞机回上海。早上在科勒镇参观了设计中心，之后去了科勒艺术中心。展品和展馆非常精彩，特别是用了很多很多很多“纸”做成的装置，相信艺术家要耗时近一年才能完成，对我的设计周展览有不一样的启发，“相信艺术，相信生活，相信韦格斯杨”，这或许可以作为之后艺术定位的口号。

向科勒学习，将艺术引入“生产”，这样可以提升价值和加深印象。

中午在科勒的小木屋吃饭，环境一流，但出品不合味，自认为很差。

经过四个多小时的车程回到芝加哥的希尔顿酒店，住了第三次。在自由活动的时间，还是选择看看名店和酒店。

在著名的密歇根大道，名牌店六七点已渐渐关门，只能看外观！幸亏酒店永远是开门的，沿途都是经典的品牌，看了很多家：Westin，Drake（Hilton的古典品牌），很古典，还有柏悦酒店。

最后选择了在柏悦吃饭，很现代的经典设计。大开眼界。幸亏遇到同团的冯先生，可以一起吃饭了。这里很旺，因为恰逢周末，餐厅没位，只能坐酒廊，也很如愿。正如帅哥说的，这是这次美国之行“最像样”的一餐。那是肯定的，品牌就是保障（正如科勒一样），我们也希望做到这样。

餐后慕名寻找芝加哥唯一一间米其林三星的餐厅：Alinen。很难找，就是一间私人屋，很难看。不知道出品会怎样，相信只是将其艺术化。坦白说，开始不相信米其林的标准了。主厨才37岁，相信只是玩技巧，吸引眼球，就像设计一样，相信很快就会不行的。我相信的是如科勒一样的“老企业”，沉淀最重要！

2013.4.26

It is the last day when I have a journey in U.S. As I have to go to the airport in the early of tomorrow, I flew back to shanghai at about 10 o'clock. In the morning I visited the design center in Kohler Town, then I went to Kohler Art Center. The products on display and exhibition is extraordinary wonderful, especially the device is made from hundreds of thousands of "paper". I believe that it should take the artist nearly one year to finish this. From that, I got the inspiration for design. "I believe in art, believe in life, believe in Weges Yang", which maybe can become the slogan and our position for art in coming future.

Learn from Kohler, introduce art into "production", thus helps enhance the value and impression.

In the afternoon, I had lunched in the small wooden house in Kohler, very good environment, but the product I think it is not tasty and personally I think it is very bad.

More than 4 hours' ride and back to Hilton Hotel in Chicago, this is the third time I lived here. It is the free time, and I choose to have a look at the famous brand and hotel.

On the famous Michigan Road, famous brand store will be closed at 6 or 7 o'clock. I can just see the outer appearance! Fortunately the hotel is always open, along the way are all the classic brands. I have visited some brands like Westin, Drake (Hilton classical brand), very classic, Park Hyatt Hotel.

Finally I still choose Park Hyatt Hotel for dinner. It has modern and classic design, which makes me surprise. Thanks to meet with Mr. Feng, the same tourist group with me, we had dinner together. It is quite a lot of people there for dinner as it was the weekend and no more seats available. So we had to sit on the wine corridor. Just one handsome guy said, this was the most decent dinner we had for this time's U.S. Journey. Surely, brand always let me have sense of safety (Just like Kohler), and we hope to do like this.

After dinner, I was looking for one canteen called Alinen, which is the only one Michelin 3 Star canteen in Chicago. It is difficult to look for it as it looks like one private house. It's not good-looking. I don't know how the taste will be. I believed that it was just the artistic way. To be frank, I never believed in the Standard of Michelin from the very beginning. The main chef was only 37 years old. I believed that he was just playing skills to attract the eyeball. It is just like design, and I believed that he won't be popular very soon. I believed that like the "old enterprise" Kohler. I think that precipitation is the most important!

April 26th, 2013

3

MICHIGAN AVENUE
4

5

3~5 / 繁华的密歇根大道，酒店林立，这可是经典的老牌酒店（THE DRAKE）。

6 / 柏悦酒店经典而时尚的大堂，矛盾吧！

7 / 周末，柏悦酒店的餐厅人头涌动，我们只能“占据”酒廊的一角，尽情享受午餐。

8 / 酒店一间接一间,柏悦与半岛相邻，我们吃饭时就能欣赏窗外美景。

9 / 晚餐后，找到芝加哥唯一一间米其林三星（Alinea）的餐厅看看。

10 / 黑麻麻的餐厅入口通道，挂满了“乱七八糟”的小玩意儿。

SULA

感谢以下人

每次出书，会惊动很多人：有帮忙整理乱七八糟的原稿；收集、核准酒店的各种资讯；初步编辑、排版；协助翻译及校对；联系出版事宜的一批同事们；有支持我在此"耗费"大量时间的朋友、家人；有一齐去体验和讨论酒店的同事们、同行们；有费了不少眼神为书写序的人：资深的"大内高手"，酒店设计的翘楚及设计界的前辈；更有耐心让我们折腾的出版社的朋友，每年一次的合作，"习以为常"了。

谢谢，我身边的人！

2016年9月18日

Acknowledgement

感谢的人

每次出书会惊动很多人：有帮忙整理乱七八糟的原稿；收集、核准酒店的各种资讯；初步编辑、排版；协助翻译及核对；联系出版事宜的一批同事们；有支持我在此“耗费”大量时间的朋友、家人；有一起去体验和讨论酒店的同事们、同行们；有费了不少眼神为书写序的人；资深的“大内高手”，酒店设计的翘楚及设计界的前辈；更有乐意让我们折腾的出版社的朋友，每年一次的合作，“习以为常”了。

谢谢，我身边的人!

2014年8月18日

Each time many persons were busy for the book's publication: someone helped to sort the messy original hand draft; someone gathered and examined all kinds of information for the hotels; someone edited and set up the type; someone helped to translate and check; and some colleagues contacted the publisher for me. There are friends and family who support me in "wasting" so much time for the book. There are colleagues and designers who experienced and discussed about hotel with me. There are also some persons who spent much energy for writing preface of the book: the expert interior designer, the expert hotel designer and the senior in design industry. There are also some staffs from publisher who accepted our torment for the book. We cooperate once a year, they have got used to it.

Thanks, people around me!

August 18th, 2014